KU-796-506

Physical Properties of Polymers

UNIVERSITY of STRATHCLYDE
FLECK
18 OCT 1985
LIBRARY
LIBRARIES

8/10/85

Physical Properties of Polymers

James E. Mark
Adi Eisenberg
William W. Graessley
Leo Mandelkern
Jack L. Koenig

American
Chemical Society

Washington, D.C.
1984

Physical
Properties
of Polymers

Includes bibliographies and index.
1. Polymers and polymerization. 2. Chemistry,
physical and theoretical.

I. Mark, James E., 1943-   . II. American Chemical
Society.

TA455.P58474   1984   620.1'92   84-10958
ISBN 0-8412-0851-4
ISBN 0-8412-0857-3 (pbk.)

Copyright © 1984
American Chemical
Society
Printed in the United
States of America

All Rights Reserved. The appearance of the code at the bottom
of the first page of each chapter in this volume indicates the
copyright owner's consent that reprographic copies of the
chapter may be made for personal or internal use or for the
personal or internal use of specific clients. This consent is
given on the condition, however, that the copier pay the stated
per copy fee through the Copyright Clearance Center, Inc., 21
Congress Street, Salem, MA 01970, for copying beyond that
permitted by Sections 107 or 108 of the U.S. Copyright Law.
This consent does not extend to copying or transmission by
any means—graphic or electronic—for any other purpose,
such as for general distribution, for advertising or
promotional purposes, for creating a new collective work, for
resale, or for information storage and retrieval systems. The
copying fee for each chapter is indicated in the code at the
bottom of the first page of the chapter.

The citation of trade names and/or names of manufacturers in
this publication is not to be construed as an endorsement or as
approval by ACS of the commercial products or services
referenced herein; nor should the mere reference herein to any
drawing, specification, chemical process, or other data be
regarded as a license or as a conveyance of any right or
permission, to the holder, reader, or any other person or
corporation, to manufacture, reproduce, use, or sell any
patented invention or copyrighted work that may in any way
be related thereto. Registered names, trademarks, etc., used in
this publication, even without specific indication thereof, are
not to be considered unprotected by law.

Second printing, 1985

D
620.192
PHY

About the Authors

James E. Mark

James E. Mark received his B.S. degree in 1957 in chemistry from Wilkes College and his Ph.D. degree in 1962 in physical chemistry from the University of Pennsylvania. After serving as a postdoctoral fellow from 1962 to 1964 at Stanford University under Professor Paul J. Flory, Dr. Mark served as assistant professor of chemistry at the Polytechnic Institute of Brooklyn from 1964 to 1967. He then moved to the University of Michigan, where he became a full professor in 1972. In 1977, he assumed the positions of professor of chemistry, chairman of the physical chemistry division, and director of the Polymer Research Center at the University of Cincinnati.

Dr. Mark's research interests pertain to the physical chemistry of polymers, including configuration-dependent properties, conformational energies of chain molecules, the elasticity of polymer networks, tin-containing polymers, and polymer-coated electrodes. Dr. Mark has lectured extensively in polymer chemistry and has published more than 150 papers. He is a fellow of both the New York Academy of Sciences and the American Physical Society.

Adi Eisenberg

Adi Eisenberg received his B.S. degree from Worcester Polytechnic Institute and his M.S. and Ph.D. from Princeton University. From 1962 to 1967, he was an assistant professor at the University of California at Los Angeles. From 1967 to 1974, he was an associate professor at McGill University. He was a visiting professor in 1973 and 1974 at the Weizmann Institute in Rehovot, Israel, and at Kyoto University in Japan. Since 1975, he has been a full professor at McGill University. His consulting has included such companies as the Jet Propulsion Laboratory in California, Owens Illinois in Ohio, and GTE in Sylvania, Mass. He has worked on more than 110 articles and patents in the polymer field and four books on ion-containing polymers. He is a member of the editorial advisory board of *Macromolecular Reviews, Journal of Polymer Science, Polymer Physics Edition,* and *Applied Physics Communications* and a member of the Advisory Committee for the Institute for Amorphous Studies in Michigan.

William W. Graessley

William W. Graessley received undergraduate degrees in chemistry and chemical engineering at the University of Michigan. As a National Science Foundation graduate fellow he obtained his Ph.D. at the same institution in 1960. After spending four years with Air Reduction Company, he joined the faculty of Northwestern University, eventually becoming Walter P. Murphy Professor of Chemical Engineering and Materials Science and Engineering. He was awarded the Bingham Medal of the Society of Rheology in 1979. He returned to industry in 1982 as senior scientific advisor in the corporate research laboratories of Exxon Research and Engineering Company.

Leo Mandelkern

Leo Mandelkern received his undergraduate degree from Cornell University in 1942. After serving with the armed forces he returned to Cornell and received his Ph.D. in 1949. He remained at Cornell in a postdoctoral capacity until 1952. Dr. Mandelkern was a member of the staff of the National Bureau of Standards from 1952 to 1962. From 1962 to the present he has been a professor of chemistry and biophysics at the Florida State University. In 1958 he received the Arthur S. Fleming Award. The American Chemical Society Award in Polymer Chemistry was bestowed upon him in 1975. He has been or is a member of the editorial boards of the *Journal of the American Chemical Society,* the *Journal of Polymer Science,* and *Macromolecules.*

Jack L. Koenig

Jack L. Koenig received his Ph.D. in 1960 from the University of Nebraska after doing his research in theoretical spectroscopy. In 1973 Dr. Koenig joined the National Science Foundation. He is presently a professor of macromolecular science and physical chemistry at Case Western Reserve University. He is active in spectroscopic research and is the director of the molecular spectroscopy laboratory. His interests include Raman spectroscopy, Fourier transform infrared spectroscopy, and solid state NMR. He is well known for his basic work in spectroscopic characterization of polymeric materials and has more than 300 publications to his credit.

Contents

Preface

The purpose of this book is to provide information on recent advances that involve the physical chemistry of polymers and that are of importance with regard to the utilization of polymeric materials. The five chapters cover rubberlike elasticity, the glassy, viscoelastic, and crystalline states in polymers, and polymer spectroscopy. The general structure of each chapter is an introduction to basic concepts, detailed descriptions of current topics of importance, comments on unsolved problems, and projections with regard to future research.

The material presented is derived from an American Chemical Society short course developed through a series of oral, audio, and teleconference presentations. It is hoped that the present format will make this material more accessible to the polymer community and that it will help to bring the reader up to date on important aspects of the subject areas covered.

James E. Mark
University of Cincinnati
Cincinnati, OH 45221

The Rubber Elastic State

James E. Mark

BASIC CONCEPTS

The most useful way to begin an article on rubberlike elasticity is to define it, and then to discuss what types of materials can exhibit this very unusual behavior. Accordingly, rubber elasticity may be operationally defined as very large deformability with essentially complete recoverability. In order for a material to exhibit this type of elasticity, three molecular requirements must be met: (i) the material must consist of polymeric chains, (ii) the chains must have a high degree of flexibility, and (iii) the chains must be joined into a network structure (1-3).

The first requirement arises from the fact that the molecules in a rubber or elastomeric material must be able to alter their arrangements and extensions in space dramatically in response to an imposed stress, and only a long-chain molecule has the required very large number of spatial arrangements of very different extensions. This versatility is illustrated in Figure 1, (1), which depicts a two-dimensional projection of a random spatial arrangement of a relatively short polyethylene chain in the amorphous state. The spatial configuration shown was computer-generated, in as realistic a manner as possible. The correct bond lengths and bond angles were employed, as was the known preference for trans rotational states about the skeletal bonds in any n-alkane molecule. A final feature taken into account is the fact that rotational states are interdependent; what one rotatable skeletal bond does, depends on what the

0851/84/0001$14.10/1
© 1984 American Chemical Society

adjoining skeletal bonds are doing. One important feature of this typical configuration is the relatively high spatial extension of some parts of the chain. This is due to the preference for the trans conformations, already mentioned, which are essentially planar zig-zag and thus of high extension. The second important feature is the fact that in spite of these preferences, many sections of the chain are quite compact. Thus, the chain extension (as measured by the end-to-end separation) is quite small. For even such a short chain, the extension could be increased approximately four-fold by simple rotations about skeletal bonds, without any need for distortions of bond angles or increases in bond lengths.

The second characteristic required for rubberlike elasticity specifies that the different spatial arrangements be accessible, i.e., changes in these arrangements should not be hindered by constraints as might result from inherent rigidity of the chains, extensive chain crystallization, or the very high viscosity characteristic of the glassy state.

The last characteristic cited is required in order to obtain the elastomeric recoverability. It is obtained by joining together or "cross-linking" pairs of segments, approximately one out of a hundred, thereby preventing stretched polymer chains from irreversibly sliding by one another. The network structure thus obtained is illustrated in Figure 2 (1), in which the cross-links may be either chemical bonds (as would occur in sulfur-vulcanized natural rubber) or physical aggregates, for example the small crystallites in a partially crystalline polymer or the glassy domains in a multi-phase block copolymer (1-3).

THE ORIGIN OF THE ELASTIC RE-TRACTIVE FORCE

The molecular origin of the elastic force f exhibited by a deformed elastomeric network can be elucidated through thermoelastic experiments, which involve the temperature dependence of either the force at constant length L or the length at constant force (1,2). Consider first a thin metal strip stretched with a weight W to a point short of that giving permanent deformation, as is shown in Figure 3 (1). Increase in temperature (at constant force) would increase the length of the stretched strip in what would be considered the "usual" behavior. Exactly the opposite, a shrinkage, is observed in the case of a stretched elastomer! For purposes of comparison, the result observed for a

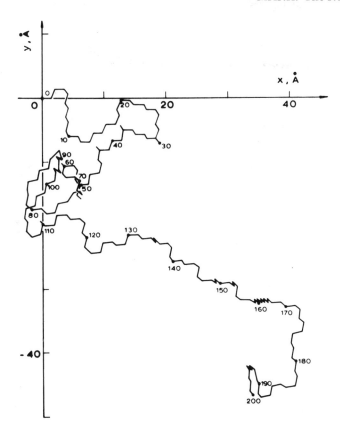

Figure 1. A two-dimensional projection of an n-alkane chain having 200 skeletal bonds (1).

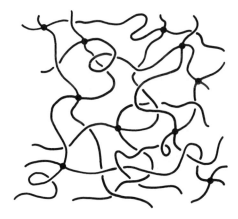

Figure 2. Schematic sketch of a typical elastomeric network (1).

gas at constant pressure is included in the figure. Raising its temperature would of course cause an increase in volume V.

The explanation for these observations is given in Figure 4.(1). The primary effect of stretching the metal is the increase ΔE in energy caused by changing the distance d of separation between the metal atoms. The stretched strip retracts to its original dimension upon removal of the force since this is associated with a decrease in energy. Similarly, heating the strip at constant force causes the usual expansion arising from increased oscillations about the minimum in the asymmetric potential energy curve. In the case of the elastomer, however, the major effect of the deformation is the stretching out of the network chains, which substantially reduces their entropy (1-3). Thus, the retractive force arises primarily from the tendency of the system to increase its entropy toward the (maximum) value it had in the undeformed state. Increase in temperature increases the chaotic molecular motions of the chains and thus increases the tendency toward the more random state. As a result there is a decrease in length at constant force, or an increase in force at constant length. This is strikingly similar to the behavior of a compressed gas, in which the extent of deformation is given by the reciprocal volume 1/V. The pressure of the gas is largely entropically derived, with increase in deformation (i.e., increase in 1/V) also corresponding to a decrease in entropy. Heating the gas increases the driving force toward the state of maximum entropy (infinite volume or zero deformation). Thus, increasing the temperature increases the volume at constant pressure, or increases the pressure at constant volume.

This suprising analogy between a gas and an elastomer (which is a condensed phase) carries over into the expressions for the work dw of deformation. In the case of a gas dw is of course -pdV. For an elastomer, however, this term is essentially negligible since network elongation takes place at very nearly constant volume (1,2). The corresponding work term now becomes +fdL, where the difference in sign is due to the fact that positive w corresponds to a decrease in volume of a gas but an increase in length of an elastomer. Similarly, adiabatically stretching an elastomer increases its temperature in the same way that adiabatically compressing a gas (for example in a diesel engine) will increase its temperature. The basic point here is the fact that the retractive

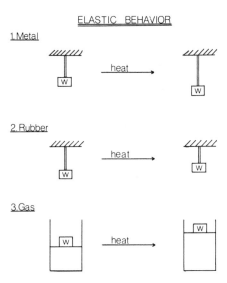

Figure 3. Results of thermoelastic experiments carried out on a typical metal, rubber, and gas (1).

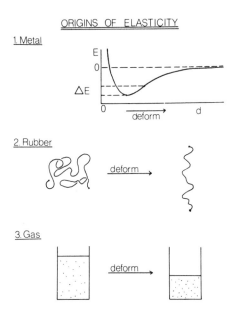

Figure 4. Sketches explaining the observations described in Figure 3 in terms of the molecular origin of the elastic force or pressure (1).

force of an elastomer and the pressure of a gas are both primarily entropically derived and, as a result, the thermodynamic and molecular descriptions of these otherwise dissimilar systems are very closely related.

Figure 5 presents a historical summary of some important qualitative concepts in the area of rubberlike elasticity. The thermoelastic experiments described above were first carried out many years ago, back in 1859. This was, in fact, only a few years after the entropy was introduced as a concept in thermodynamics in general! A molecular interpretation of the fact that rubberlike elasticity is primarily entropic in origin had to await Hermann Staudinger's much more recent demonstration that polymers were covalently bonded molecules, and not some type of association complex best studied by the colloid chemists (2). Another important experimental fact relevant to the development of these molecular ideas was the fact that deformations of rubberlike materials occurred essentially at constant volume, so long as crystallization was not induced (2). (In this sense, the deformation of an elastomer and a gas are very different). Werner Kuhn used this observed constancy in volume to point out that the changes in entropy must therefore involve changes in orientation or configuration of the network chains. These basic qualitative ideas are shown in the sketch in Figure 5, where the arrows represent some typical end-to-end vectors of the network chains.

In the 1930's, Kuhn, Eugene Guth, and Herman Mark first began to develop quantitative theories based on this idea that the network chains undergo configurational changes, by skeletal bond rotations, in response to an imposed stress (2,3). These theories, and some of their modern-day refinements, are described in the following Section.

THE ELASTIC
FREE ENERGY AND
ELASTIC EQUA-
TIONS OF STATE

Any molecular theory of rubberlike elasticity is based on a chain distribution function, which gives the probability of any end-to-end separation r. The characteristics of this type of distribution function are given in Figure 6 (2). What is required is a function that answers the question "If a chain starts at the origin of the coordinate system shown, what is the probability that the other end will be in a infinitesimal volume dxdydz around some specified values of x, y, and z?"

I.   Thermodynamic Origin of the Force f

        Thermoelastic measurements (f-L-T) indicated
        f was entropic rather than energetic
        (Joule, Lord Kelvin, 1859).

II.   Molecular Picture

    A.   Polymers are long,
        covalently-bonded
        chains (Staudinger,
        ∿1920).

    B.   Orientation of chains
        is origin of ΔS (Meyer, 1932).

    C.   ΔS and f are due to
        configurational changes,
        i.e., rotations about
        skeletal bonds (Kuhn,
        Guth, Mark, 1934-39).

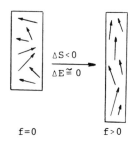

$\Delta S < 0$

$\Delta E \cong 0$

$f = 0$ $f > 0$

Figure 5. A summary of some important qualitative concepts in rubberlike elasticity.

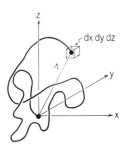

Figure 6. A spatial configuration of a polymer chain, with some quantities used in the distribution function for the end-to-distance r. (Reproduced with permission from Ref. 2. Copyright 1953, Cornell University Press.)

The simplest molecular theories of rubberlike elasticity are based on the Gaussian distribution function

$$w(r) = \left(\frac{3}{2\pi <r^2>_o}\right)^{3/2} \exp\left(-\frac{3\ r^2}{2\ <r^2>_o}\right) \qquad (1)$$

for the end-to-end separations of the network chains (i.e., chain sequences extending from one cross-link to another) (1-3). In this equation, $<r^2>_o$ represents the dimensions of the free chains as unperturbed by excluded volume effects (2). These excluded volume interactions arise from the spatial requirements of the atoms making up the polymeric chain and are thus similar to those occurring in gases. They are more complex, however, in that they have an intramolecular as well as intermolecular origin. If present, they increase the dimensions of a polymer chain in the same way they can increase the pressure of a gas. The Gaussian distribution function in which $<r^2>_o$ resides is applied to the network chains in both the stretched and unstretched states. The Helmholtz free energy of such a chain is given by the simple variant of the Boltzmann relationship shown in the first part of the equation

$$F(T) = -kT\ \ln\ w(r) = C(T) + \left(\frac{3kT}{2\ <r^2>_o}\right)r^2 \qquad (2)$$

where $C(T)$ is a constant at a specified absolute temperature T. Consider now the process of stretching a network chain from its random undeformed state with r components of x,y,z, to the deformed state with r components of $\alpha_x x$, $\alpha_y y$, $\alpha_z z$, (where the α's are molecular deformation ratios). The free energy change for a single network chain is then simply

$$\Delta F = \left(\frac{3kT}{2\ <r^2>_o}\right)\left[(\alpha_x^2 x^2 + \alpha_y^2 y^2 + \alpha_z^2 z^2) - \right.$$

$$\left. (x^2 + y^2 + z^2)\right] \qquad (3)$$

Since the elastic response is essentially entirely intramolecular (1-3), the free energy change for ν network chains is just ν times the above result

$$\Delta F = \left(\frac{3\nu kT}{2\ <r^2>_o}\right)\left[(\alpha_x^2 - 1)<x^2> + (\alpha_y^2 - 1)<y^2> + \right.$$

$$\left. (\alpha_z^2 - 1)<z^2>\right] \qquad (4)$$

where the brackets around x^2, y^2, and z^2 specify their values when averaged over the ν chains. It has generally been assumed that the strain-induced displacements of the cross-links or junction points are affine (i.e. linear) in the macroscopic strain. In this case, the deformation ratios are obtained directly from the dimensions of the sample in the strained state and in the initial unstrained state

$$\alpha_x = L_x/L_{xi}, \; \alpha_y = L_y/L_{yi}, \; \alpha_z = L_z/L_{zi} \qquad (5)$$

The dimensions of the cross-linked chains in the undeformed state are given by the pythagorean theorem

$$\langle r^2 \rangle_i = \langle x^2 \rangle + \langle y^2 \rangle + \langle z^2 \rangle \qquad (6)$$

Also, the isotropy of the undeformed state requires that the average values of x^2, y^2, and z^2, be the same, i.e.

$$\langle x^2 \rangle = \langle y^2 \rangle = \langle z^2 \rangle \qquad (7)$$

Thus, the chain dimensions are given by

$$\langle r^2 \rangle_i = 3\langle x^2 \rangle = 3\langle y^2 \rangle = 3\langle z^2 \rangle \qquad (8)$$

and the elastic free energy of deformation by

$$\Delta F = (\frac{\nu kT}{2}) \left[\frac{\langle r^2 \rangle_i}{\langle r^2 \rangle_o} \right] (\alpha_x^2 + \alpha_y^2 +$$

$$\alpha_z^2 - 3) \qquad (9)$$

In the simplest theories (1-3), $\langle r^2 \rangle_i$ is assumed to be identical to $\langle r^2 \rangle_o$; i.e., it is assumed that the cross-links do not significantly change the chain dimensions from their unperturbed values. Equation (9) may then be approximated by

$$\Delta F \simeq (\frac{\nu kT}{2})(\alpha_x^2 + \alpha_y^2 + \alpha_z^2 - 3) \qquad (10)$$

Equations (9) and (10) are basic to the molecular theories of rubberlike elasticity and can be used to obtain the elastic equations of state for any type of deformation (1-3), i.e., the equations inter-relating the stress, strain, temperature, and number or number density of network chains. Their application is best illustrated in the case of elongation, which is the type of deformation used in the great majority of experimental studies (1-3). This deformation occurs at essentially constant volume and thus a network stretched by the amount of $\alpha_x = \alpha > 1$ would have its perpendicular dimensions compressed by the amounts

$$\alpha_y = \alpha_z = \alpha^{-\frac{1}{2}} < 1 \qquad (11)$$

Accordingly, for elongation, one obtains the first part of the equation

$$\Delta F = (\frac{\nu kT}{2})(\alpha^2 + 2\alpha^{-1} - 3) = fdL \qquad (12)$$

Since the Helmholtz free energy is the "work function" and the work of deformation is fdL (where $L = \alpha L_i$), the elastic force may be obtained by differentiating Equation 12, to obtain

$$f = (\partial \Delta F/\partial L)_{T,V} = (\frac{\nu kT}{L_i})(\alpha - \alpha^{-2}) \qquad (13)$$

The nominal stress $f^* \equiv f/A^*$, where A^* is the undeformed cross-sectional area, is then given by

$$f^* \equiv f/A^* = (\nu kT/V)(\alpha - \alpha^{-2}) \qquad (14)$$

where ν/V is the density of network chains, i.e., their number per unit volume V, which is L_iA^*.

The elastic equation of state in the form given in Equation (14) is strikingly similar to the molecular form of the equation of state for an ideal gas

$$p = NkT(1/V) \qquad (15)$$

where the stress has replaced the pressure and the number density of network chains has replaced the number of gas molecules. Similarly, since the stress was assumed to be entirely entropic in origin, f^* is predicted to be directly proportional to T at constant α (and V), as is the pressure of the ideal gas at constant 1/V. The strain function $(\alpha - \alpha^{-2})$ is somewhat more complicated than is 1/V since the near incompressibility of the elastomeric network superposes compressive effects (given by $-\alpha^{-2}$) on the simple elongation (α) being applied to the system.

Also frequently employed in elasticity studies is the "reduced stress" or modulus defined in the first part of the equation

$$\left[f^*\right] \equiv f^*v_2^{1/3}/(\alpha - \alpha^{-2}) = \nu kT/V \qquad (16)$$

Its definition includes a factor which makes it

applicable to networks which are swollen, which is frequently done to facilitate the approach to elastic equilibrium. The factor, which is the cube root of the volume fraction of polymer in the network, takes into account the fact that a swollen network has fewer chains passing through unit cross-sectional area, and that the chains are stretched due to the presence of diluent (2).

Recent work suggests that the elongation of a network is generally not affine (4-7). Particularly at higher deformations, it is thought that junction fluctuations diminish the modulus by the factor $A_\phi < 1$ in the modified equation

$$\left[f^*\right] = A_\phi \nu kT/V \qquad (17)$$

In the limit of the very non-affine deformation which would be exhibited by a "phantom" network (where the chains are portrayed as being able to transect one another) A_ϕ is given by

$$A_\phi = 1 - 2/\phi \qquad (18)$$

where ϕ is the cross-link functionality (the number of chains emanating from a network cross-link) (5-7).

The basic elastic equation of state, and its recent refinements, are discussed further in subsequent Sections.

SOME EXPERIMENTAL DETAILS

The apparatus typically used to measure the force required to give a specified elongation in a rubberlike material is very simple, as can be seen from its schematic description in Figure 7 (1). The elastomeric strip is mounted between two clamps, the lower one fixed and the upper one attached to a movable force gauge. A recorder is used to monitor the output of the gauge as a function of time in order to obtain equilibrium values of the force suitable for comparisons with theory. The sample is generally protected with an inert atmosphere, such as nitrogen, to prevent degradation, particularly in the case of measurements carried out at elevated temperatures. Both the sample cell and surrounding constant-temperature bath are glass, thus permitting use of a cathetometer or travelling microscope to obtain values of the strain by measurements of the distance between two lines marked on the central portion of the test sample.

THE DEPENDENCE OF THE STRESS ON ELONGATION

A typical stress-strain isotherm obtained on a strip of cross-linked natural rubber as described above is shown in Figure 8 (1,3). The units for the force are generally Newtons, and the curves obtained are usually checked for reversibility. In this type of representation, the area under the curve is frequently of considerable interest since it is proportional to the work of deformation w = $\int f dL$. Its value up to the rupture point is thus a measure of the toughness of the material.

The initial part of the stress-strain isotherm shown in Figure 8 is of the expected form in that f^* approaches linearity with α as α becomes sufficiently large to make the α^{-2} term in Equation (14) negligibly small. The large increase in f^* at high deformation in the case of natural rubber is due largely if not entirely to strain-induced crystallization, as is described in the section on non-Gaussian effects. The melting point of the polymer is inversely proportional to the entropy of fusion, which is significantly diminished when the chains in the amorphous network remain stretched out because of the applied deformation. The melting point is thereby increased and it is in this sense that the stretching "induces" the crystallization of some of the network chains. The effect is qualitatively similar to the increase in melting point generally observed upon increase in pressure on a low molecular weight substance in the crystalline state. In any case, the crystallites thus formed act as physical cross-links, increasing the modulus of the network. The properties of both crystallizable and non-crystallizable networks at high elongations are discussed further in subsequent Sections.

Additional deviations from theory are found in the region of moderate deformation upon examination of the usual plots of modulus against reciprocal elongation (3,8). Although Equation (16) predicts the modulus to be independent of elongation, it generally decreases significantly upon increase in α. Typical results, obtained on swollen and unswollen networks of natural rubber, are shown in Figure 9 (8). The intercepts and slopes of such linear plots are generally called the Mooney-Rivlin constants $2C_1$ and $2C_2$, respectively. It is important to note that the slope $2C_2$, a measure of the discrepancy from the predicted behavior, decreases to an essentially negligible value as the degree of swelling of the network increases. In the more refined molecular theories of rubberlike elasticity (4-7) this decrease is explained by the gradual increase in the non-affineness of the deformation as the elongation increases, as is

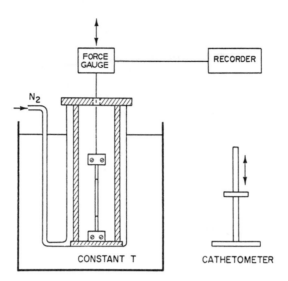

Figure 7. Apparatus for carrying out stress-strain measure-ments on an elastomer in elongation (1).

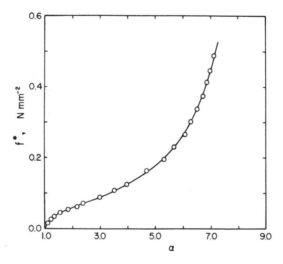

Figure 8. Stress-elongation curve for natural rubber in the vicinity of room temperature (1, 3).

shown schematically in Figure 10. In the case of the two limits, the affine deformation and the nonaffine deformation in the phantom network limit, the reduced stress should be independent of α. The value for the phantom limit should be reduced, however, by the factor $(1 - 2/\phi)$ in the case of a ϕ-functional network, as is illustrated for the case $\phi = 4$. The experimentally observed decreases in reduced stress with increasing α are shown as the heavier portion of the theoretical curve. In these theories, the degree of entangling around the cross-links is of primary importance, since this will determine the firmness with which the cross-links are embedded in the network structure. This type of chain-cross-link entangling is illustrated in Figure 11 (9). For a typical degree of cross-linking, there are 50-100 cross-links closer to a given cross-link than those directly joined to it through a single network chain. The configurational domains thus generally severely overlap. The degree of overlapping is a measure of the firmness with which the cross-links are embedded, and thus of the extent to which the idealized, affine deformation is approached. Stretching out the network chains decreases this degree of entangling, thereby permitting increased cross-link fluctuations. The modulus thus decreases, approaching the value predicted for a phantom network, where entangling is impossible and cross-link fluctuations are unimpeded. This concept also explains the essentially constant modulus at high degrees of swelling illustrated in Figure 9. Large amounts of diluent "loosen" the cross-links so that the deformation is highly non-affine even at low deformations, and thus the modulus changes relatively little upon increase in elongation.

THE DEPENDENCE OF STRESS ON TEMPERATURE

As already mentioned, the assumption of a purely entropic elasticity leads to the prediction, Equation (14), that the stress should be directly proportional to the absolute temperature at constant α (and V). The extent to which there are deviations from this direct proportionality may therefore be used as a measure of the thermodynamic non-ideality of an elastomer (10-13). In fact, the definition of ideality for an elastomer is that the energetic contribution f_e to the elastic force f be zero. This quantity is defined by

$$f_e \equiv (\partial E/\partial L)_{V,T} \tag{19}$$

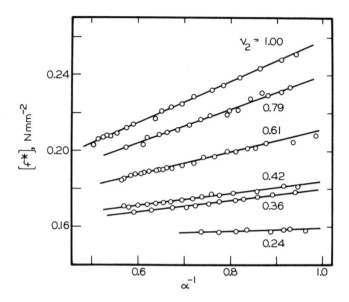

Figure 9. The modulus shown as a function or reciprocal elongation as suggested by the semi-empirical Mooney-Rivlin equation $[f^*] = 2C_1 + 2C_2\alpha^{-1}$ (3, 8). The elastomer is natural rubber, both unswollen and swollen with n-decane (8). Each isotherm is labelled with the volume fraction of polymer in the network.

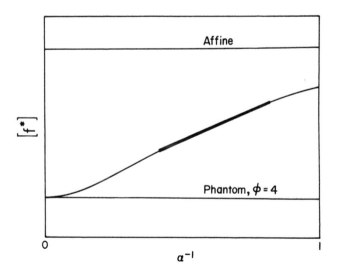

Figure 10. A schematic diagram qualitatively showing theoretical predictions (5-7) for the reduced stress as a function of a reciprocal elongation α^{-1}.

and this definition is obviously a close parallel to the requirement that $(\partial E/\partial V)_T$ be zero for ideality in a gas.

Force-temperature ("thermoelastic") measurements may therefore be used to obtain experimental values of the fraction f_e/f of the force which is energetic in origin. Such experiments at constant volume are the most direct, and can be interpreted through use of the relationship

$$f_e/f = -T\left[\partial \ln(f/T)/\partial T\right]_{V,L} \tag{20}$$

Since, however, it is very difficult to maintain constant volume in these experiments, they are usually carried out at constant pressure instead. They are then interpreted using the equation

$$f_e/f = -T\left[\partial \ln(f/T)/\partial T\right]_{p,L} - \beta T/(\alpha^3 - 1) \tag{21}$$

in which β is the thermal expansion coefficient of the network. This relationship was obtained by using the Gaussian elastic equation of state to correct the data to constant pressure (10-13).

The energy changes are intramolecular (10-13) and arise from transitions of the chains from one spatial configuration to another [since different configurations generally correspond to different intramolecular energies (14)]. They are thus obviously related to the temperature coefficient of the unperturbed dimensions, the quantitative relationship

$$f_e/f = T \; d\ln\langle r^2\rangle_0/dT \tag{22}$$

being obtained by keeping the $\langle r^2\rangle_i$ factor in Equation (9) distinct from $\langle r^2\rangle_0$. [It is interesting to note that since this type of non-ideality is intramolecular, it is not removed by diluting the chains (swelling the network) nor by increasing the lengths of the network chains (decreasing the degree of cross-linking). In this respect, elastomers are rather different from gases, which can be made to behave ideally by decreasing the pressure to a sufficiently low value].

Some typical thermoelastic data, obtained on amorphous polyethylene, are shown in Figure 12 (11,13). Their interpretation using Equation (21) indicates that the energetic contribution to the elastic force is large and negative. These results on polyethylene (11) may be understood using the information given in Figure 13. The preferred (lowest energy) conformation of the chain is the all-trans form, since gauche states (at rotational

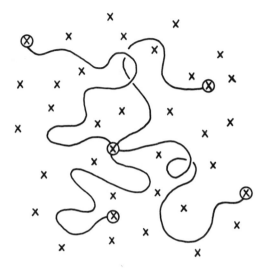

Figure 11. *Typical configurations of four chains emanating from a tetrafunctional cross-link in a polymer network prepared in the undiluted state (9).*

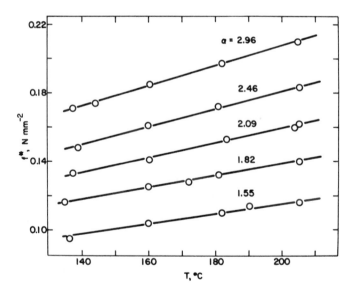

Figure 12. *Some thermoelastic data for amorphous polyethylene networks at constant length (11). Each curve is labelled with the value of the elongation at the highest temperature of measurement. (Reproduced with permission from Ref. 13. Copyright 1976, John Wiley & Sons, Inc.)*

angles of $\pm120°$) cause steric repulsions between CH_2 groups ($\underline{14}$). Since this conformation has the highest possible spatial extension, stretching a polyethylene chain requires switching some of the gauche states (which are of course present in the randomly-coiled form) to the alternative trans states ($\underline{11-14}$). These changes decrease the conformational energy and are the origin of the negative type of ideality represented in the experimental value of f_e/f. (This physical picture also explains the decrease in unperturbed dimensions upon increase in temperature. The additional thermal energy causes an increase in the number of the higher energy gauche states, which are more compact than the trans ones).

The opposite behavior is observed in the case of poly(dimethylsiloxane), as is described in Figure 14. The all-trans form is again the preferred conformation; the relatively long Si-O bonds and the unusually large Si-O-Si bond angles reduce steric repulsions in general, and the trans conformation places CH_3 side groups at distances of separation where they are strongly attractive ($\underline{12-14}$). Because of the inequality of the Si-O-Si and O-Si-O bond angles, however, this conformation is of very low spatial extension. Stretching a poly-(dimethylsiloxane) chain therefore requires an increase in the number of gauche states. Since these of higher energy, this explains the fact that deviations from ideality for these networks are found to be positive ($\underline{12-14}$).

Thermoelastic results are also used to test some of the assumptions used in the development of the molecular theories. Specifically, the results given in Table I ($\underline{13}$) indicate that the ratio f_e/f

Table I

Effect of Dilution on f_e/f

POLYMER	DILUENT	$v_2{}^a$	f_e/f
Polyethylene	None	1.00	-0.42 (±0.04)
	Diethylhexyl azelate	0.80-0.30	-0.44 (±0.10)
	$\underline{n}$-$C_{30}H_{62}$	0.50	-0.64
	$\underline{n}$-$C_{32}H_{66}{}^b$	~0.30	-0.50 (±0.06)
Natural rubber	None	1.00	0.17 (±0.03)
	$\underline{n}$-$C_{16}H_{34}$	0.98-0.34	0.18 (±0.04)
	$\underline{n}$-$C_{10}H_{22}$	0.65-0.36	0.13 (±0.01)
	Paraffin oil	0.40	0.19 (±0.02)
	Decalin ^{b}	~0.20	0.14 (±0.02)
Trans-1,4-polyisoprene	None	1.00	-0.10 (±0.05)
	Paraffin oil	0.40	-0.13 (±0.02)
	Decalin ^{b}	~0.18	-0.20 (±0.04)

^{a} Volume fraction of polymer in the network.
^{b} Swelling equilibrium.

A. RESULTS

$$\frac{f_e}{f} = T\frac{d \ln \langle r^2 \rangle_0}{dT} = -0.45$$

B. INTERPRETATION

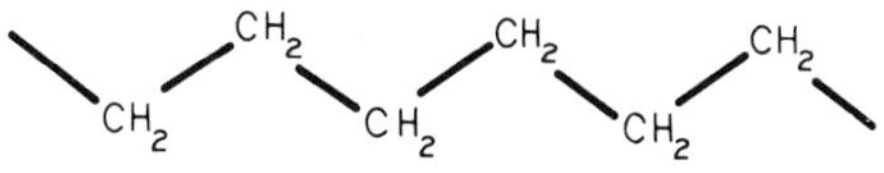

Figure 13. Thermoelastic results on (amorphous) polyethylene networks and their interpretation in terms of the preferred, all-trans conformation of the chain (1, 14).

A. Results

$$\frac{f_e}{f} = T\frac{d \ln \langle r^2 \rangle_0}{dT} = 0.25$$

B. Interpretation

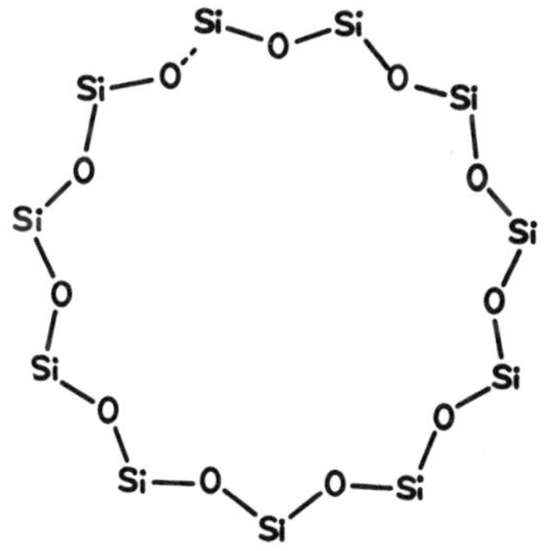

Figure 14. Thermoelastic results on poly(dimethylsiloxane) networks and their interpretation in terms of the preferred, all-trans conformation of the chain (1, 14). For purposes of clarity, the two methyl groups on each silicon atom have been deleted.

is essentially independent of degree of swelling of
the network, and this supports the postulate made
in an earlier Section that intermolecular
interactions do not contribute significantly to the
elastic force. The assumption is further supported
by the results given in Table II ([13]), which show

Table II

Comparison of d $\ln<r^2>_0/dT$ Deduced from Thermoelastic Measurements
on Networks with Values from Viscometric Measurements on Isolated Chains

POLYMER	10^3 d $\ln<r^2>_0/dT$	
	f-T	[η]-T
Polyethylene	-1.05 ($\pm$0.10)	-1.10 ($\pm$0.07)
		-1.09 ($\pm$0.04)
		-0.8 ($\pm$0.1)
Poly(n-pentene-1) Isotactic	0.34 ($\pm$0.04)	0.52 ($\pm$0.05)
Polystyrene Atactic	0.37 ($\pm$0.08)	0.56 ($\pm$0.10)
Polyisobutylene	-0.19 ($\pm$0.11)	-0.28 ($\pm$0.05)
		-0.4 ($\pm$0.1)
Polyoxyethylene	0.23 ($\pm$0.02)	0.2 ($\pm$0.2)
Polydimethylsiloxane	0.59 ($\pm$0.14)	0.52 ($\pm$0.20)

that the values of the temperature coefficient of
the unperturbed dimensions obtained from
thermoelastic experiments are in good agreement
with those obtained from viscosity-temperature
measurements on the isolated chains in dilute
solution. Since intermolecular interactions do not
affect the force, they must be independent of the
extent of the deformation and thus the spatial
configurations of the chains. This in turn
indicates that the spatial configurations must be
independent of intermolecular interactions, i.e.,
the amorphous chains must be in random, unordered
configurations, the dimensions of which should be
the unperturbed values ([2]). This conclusion has
now been amply verified, in particular by neutron
scattering studies on undiluted amorphous polymers.
These uses of thermoelastic results are summarized
in Table III ([13]).

THE DEPENDENCE OF THE STRESS ON NETWORK STRUC- TURE

Until recently, there was relatively little
reliable quantitative information on the
relationship of stress to structure, primarily
because of the uncontrolled manner in which
elastomeric networks were generally prepared ([1-3]).

Table III

Summary and Conclusions, Thermoelasticity Studies

A. Evidence

1. Values of $f_e/f = T \, d\ln \langle r^2 \rangle_0 /dT$ are independent of cross-linking conditions, degree of cross-linking, type and extent of deformation, and presence of diluent in a network.

2. Values of $d\ln \langle r^2 \rangle_0 dT$ from thermoelastic studies are in good agreement with values from $[\eta]$-T experiments.

B. Conclusions

1. Intermolecular interactions must be independent of the extent of deformation (i.e., f_e/f must be intramolecular).

2. Chains in the bulk amorphous state must be in random, unordered configurations.

Segments close together in space were linked irrespective of their locations along the chain trajectories, thus resulting in a highly random network structure in which the number and locations of the cross-links were essentially unknown. Such a structure is shown in Figure 2. New synthetic techniques are now available, however, for the preparation of "model" polymer networks of known structure (15-18). An example is the reaction shown in Figure 15, in which hydroxyl-terminated poly(dimethylsiloxane) (PDMS) chains are end-linked using tetraethyl orthosilicate. Characterizing the uncross-linked chains with respect to molecular weight M_n and molecular weight distribution, and then running the specified reaction to completion gives elastomers in which the network chains have these characteristics, in particular a molecular weight M_c between cross-links equal to M_n, and the cross-links have the functionality of the end-linking agent.

Trifunctional and tetrafunctional PDMS networks prepared in this way have been used to test the molecular theories of rubber elasticity with regard to the increase in non-affineness of the network deformation with increasing elongation. Some of these results are shown in Figure 16 (16). The ratio $2C_2/2C_1$ decreases with increase in cross-

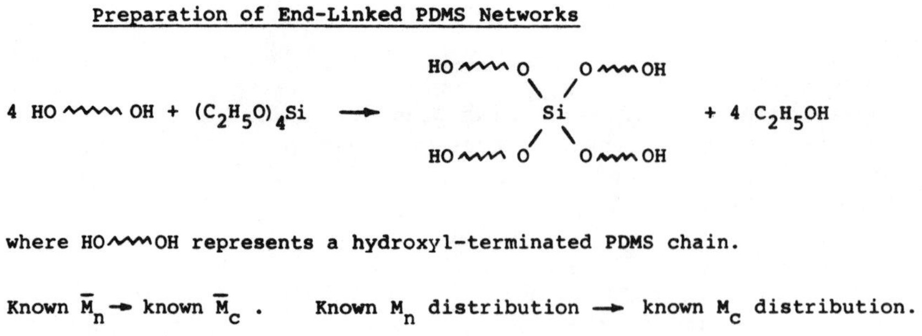

Preparation of End-Linked PDMS Networks

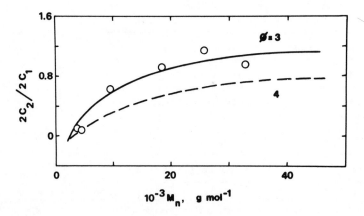

4 HO $\sim\!\!\sim\!\!\sim$ OH + $(C_2H_5O)_4Si$ $\longrightarrow$ [structure] + 4 C_2H_5OH

where HO$\sim\!\!\sim\!\!\sim$OH represents a hydroxyl-terminated PDMS chain.

Known $\bar{M}_n \rightarrow$ known $\bar{M}_c$.   Known M_n distribution $\rightarrow$ known M_c distribution.

Figure 15. A typical synthetic route for preparing elastomeric networks of known structure (1, 15).

Figure 16. Experimental data showing values of the ratio $2C_2/2C_1$, which is a measure of the increase in non-affineness of the deformation as the elongation increases. The ratio decreases with increase in junction functionality and with decrease in network chain molecular weight, as predicted by theory (5-7).

link functionality from three to four because cross-links connecting four chains are more constrained than those connecting only three. There is therefore less of a decrease in modulus brought about by the fluctuations which are enhanced at high deformation and give the deformation its non-affine character. The decrease in $2C_2/2C_1$ with decrease in network chain molecular weight is due to the fact that there is less configurational interpenetration in the case of short network chains. This decreases the firmness with which the cross-links are embedded and thus the deformation is already highly non-affine even at relatively small deformations.

A more thorough investigation of the effects of cross-link functionality requires use of the more versatile chemical reaction illustrated in Figure 17. Specifically, vinyl-terminated PDMS chains are end-linked using a multi-functional silane. In the study summarized in Figure 18 (17), this reaction was used to prepare PDMS model networks having functionalities ranging from three to eleven, with a relatively unsuccessful attempt to achieve a functionality of thirty-seven. As shown in the figure, the modulus $2C_1$ increases with increase in functionality, as expected from the increased constraints on the cross-links, and as predicted in Equations (17) and (18). Similarly, $2C_2$ and its value relative to $2C_1$ both decrease, for the reasons described in the discussion of Figure 16.

Such model networks may also be used to provide a direct test of molecular predictions of the modulus of a network of known degree of cross-linking. Before commenting further on such experiments, however, it is useful to digress briefly to establish the relationship between the three most widely used measures of the crosslink density. The first involves the number (or number of moles) ν of network chains, where a network chain is defined as one which extends from one crosslink to another. This quantity is usually expressed as the chain density ν/V, where V is the volume of the (unswollen) network (2). A second measure, directly proportional to it, is the density μ/V of crosslinks. The relationship between the number of crosslinks μ and the number of chains ν must obviously depend on the crosslink functionality. The two most important types of networks in this regard are the tetrafunctional (ϕ = 4), almost invariably obtained upon joining two segments from different chains , and the

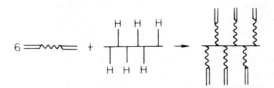

Figure 17. A typical reaction in which vinyl-terminated PDMS chains are end-linked with a multifunctional silane.

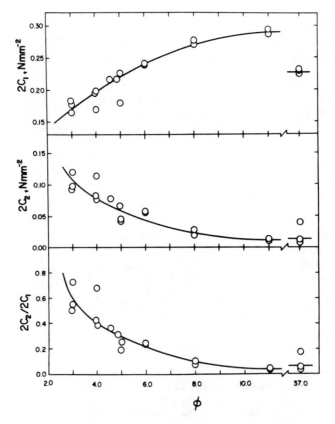

Figure 18. Experimental data showing the effect of cross-link functionality on $2C_1$ (a measure of the high deformation modulus) and $2C_2$ and $2C_2/2C_1$ (17).

trifunctional (obtained for example when forming a polyurethane network by end-linking hydroxyl-terminated chains with a triisocyanate). The relationship between μ and ν is illustrated in Figure 19 ([18]), which consists of sketches of two simple, perfect network structures, the first tetrafunctional and the second trifunctional. They are simple in the sense of having small enough values of μ and ν to be counted, and perfect in the sense of not having any dangling ends or elastically ineffective loops (chains with both ends attached to the same cross-link). As can be seen, the tetrafunctional network yields μ/ν = 4/8 or 1/2, and the trifunctional one 4/6 or 2/3. For a perfect ϕ-functional network the number $\phi\mu$ of crosslink attachment points equals the number 2ν of chain ends, thus giving the simple relationship ([2])

$$\mu = (2/\phi)\nu \qquad (23)$$

A final (inverse) measure of crosslink density is the molecular weight M_c between crosslinks. This is simply the density (ρ, g cm^{-3}) divided by the number of moles of chains (ν/V, moles cm^{-3})([2])

$$M_c = \rho/(\nu/V) \qquad (24)$$

Some experiments on model networks ([15-17]) have given values of the elastic modulus in good agreement with theory. Others have given values significantly larger than predicted, and the increases in modulus have been attributed to contributions from "permanent" chain entanglements of the type shown in Figure 20. There are disagreements, and the issue has not yet been entirely resolved. Since the relationship of modulus to structure is of such fundamental importance, there is currently a great deal of research activity in this area ([1]).

NETWORKS PREPARED UNDER UNUSUAL CONDITIONS

Two techniques which may be used to prepare networks having simpler topologies are illustrated in Figure 21 ([19,20]). Basically, they involve separating the chains prior to their cross-linking by either stretching or dissolution. After the cross-linking, the stretching force or solvent is removed and the network is studied (unswollen) with regard to its stress-strain properties in elongation. Some results obtained on PDMS networks cross-linked in solution by means of γ radiation are shown in Table IV ([20]). There is seen to be a

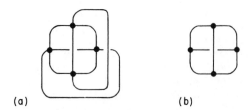

(a)   (b)

Figure 19. Sketches of some simple, perfect networks having (a) tetrafunctional and (b) trifunctional cross-links (both of which are indicated by the heavy dots). (Reproduced with permission from Ref. 18. Copyright 1982, Rubber Chem. Technol.)

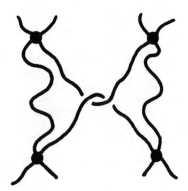

Figure 20. Sketch of a "trapped" interchain entanglement.

Table IV. PDMS Networks Compared at Approximately Constant Modulus

$v_{2,s}^a$	t_{eq}, hr	$\dfrac{[f^*]\,(\text{equil})}{[f^*]\,(\text{init})}$	$2C_2$, N mm^{-2}
1.00	0.70	0.95	0.062
0.75	0.48	0.98	0.057
0.62	0.10	0.99	0.059
0.55	0.30	0.99	0.062
0.048	0.02	1.00	0.067
0.40	0.03	1.00	0.039
0.30	0.00	1.00	0.031

[a]Volume fraction of polymer in the solution being irradiated in the cross-linking reaction.

continual decrease in the time required to reach elastic equilibrium and in the extent of stress relaxation, upon decrease in the volume fraction of polymer present during the cross-linking. Also, at higher dilutions there is a decrease in the Mooney-Rivlin $2C_2$ constant as well.

These observations are qualitatively explained in Figure 22. If a network is cross-linked in solution and the solvent then removed, the chains collapse in such a way that there is reduced overlap in their configurational domains. It is primarily in this regard, namely decreased chain-cross-link entangling, that solution-cross-linked samples have simpler topologies, with correspondingly simpler elastomeric behavior.

It is appropriate to comment at this point on the opposite sort of experiment, cross-linking a network in the undiluted state and then studying its stress-strain isotherms in the swollen state. Such a diluent might be introduced to suppress crystallization or to facilitate the approach to elastic equilibrium. Figure 23 (<u>21</u>), however,

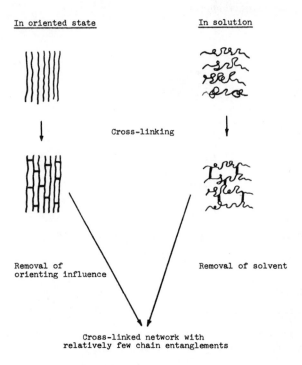

Figure 21. Two techniques that may be used to prepare networks of simpler topology (19, 20).

Figure 22. Typical configurations of four chains emanating from a tetrafunctional cross-link in a (dried) polymer network prepared in solution.

illustrates a complication which can occur in the case of networks of polar polymers at relatively high degrees of swelling. The observations is that different solvents, at the same degree of swelling, can have significantly different effects on the elastic force. This is apparently due to a "specific solvent effect" on the unperturbed dimensions, which appear in the basic relationship given in Equation (9). Although frequently observed in studies of the solution properties of uncross-linked polymers, the effect is not yet well understood. It is apparently partly due to the effect of the solvent's dielectric constant on the Coulombic interactions between parts of a chain, but probably also to solvent-polymer segment interactions that change the conformational preferences of the chain backbone (21).

As already described in Figure 8 (1,3), some (unfilled) networks show a large and rather abrupt increase in modulus at high elongations. This increase, which is further illustrated for natural rubber in Figure 24 (22,23), is very important since it corresponds to a significant toughening of the elastomer. Its molecular origin, however, has been the source of considerable controversy. It had been widely attributed to the "limited extensibility" of the network chains (24), i.e., to an inadequacy in the Gaussian distribution function. This potential inadequacy is readily evident in the exponential in Equation (1), specifically that this function does not assign a zero probability to a configuration unless its end-to-end separation r is infinite. This explanation was viewed with skepticism by some workers since the increase in modulus was generally observed only in networks which could undergo strain-induced crystallization. Such crystallization in itself could account for the increase in modulus, primarily because the crystallites thus formed would act as additional cross-links in the network.

Attempts to clarify the problem by using non-crystallizable networks (23) were not convincing since such networks were incapable of the large deformations required to distinguish between the two possible interpretations. The issue has now been resolved (24-28), however, by the use of end-linked, non-crystallizable model PDMS networks. These networks have high extensibilities, presumably because of their very low incidence of

NON-GAUSSIAN
EFFECTS

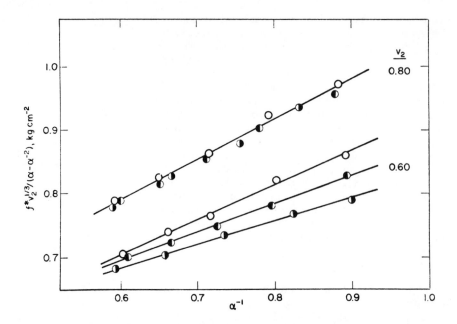

Figure 23. Stress-strain isotherms (21) for the same PDMS network swollen by different solvents, specifically dimethylsiloxane oligomer (○), n-hexadecane (◑), and n-octyl acetate (◐).

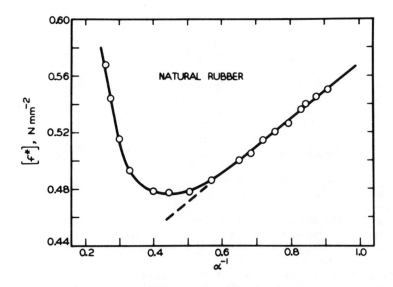

Figure 24. Stress-strain isotherm for an unfilled rubber network at 25 °C, showing the anomalous increase in modulus at high elongation (22). (Reproduced with permission from Ref. 23. Copyright 1976, John Wiley & Sons, Inc.)

dangling-chain network irregularities. They have particularly high extensibilities when they are prepared from a mixture of very short chains (around a few hundred g mol^{-1}) with relatively long chains (around 18,000 g mol^{-1}). Apparently the very short chains are important because of their limited extensibilities, and the relatively long chains because of their ability to retard the rupture process.

Some typical results on such bimodal PDMS networks are shown in Figure 25 (24). The upturns in modulus are much less pronounced than they are in crystallizable polymer networks such as natural rubber or cis-1,4-polybutadiene, and they are independent of temperature, as would be expected in the case of limited chain extensibility. For a crystallizable network, the upturns diminish and eventually disappear upon increase in temperature, as is illustrated for cis-1,4-polybutadiene in Figure 26 (25,26). Similarly, as shown in Figure 27 (24), swelling has relatively little effect on the upturns in the case of PDMS. Again, the upturns in crystallizable polymer networks disappear upon sufficient swelling, as illustrated in Figure 28 (26,27).

Two additional features in Figure 28, however, merit additional comments. First, the initiation of the strain-induced crystallization (as evidenced by departure of the isotherm from linearity) is facilitated by the presence of the low molecular weight diluent. Thus, in a sense this kinetic effect acts in opposition to the thermodynamic effect, which is primarily the suppression of the polymer melting point by the diluent. The second interesting point has to do with the decrease in the modulus prior to its increase. As shown schematically in Figure 29 (9,28), this is probably due to the fact that the crystallites are oriented along the direction of stretching, and the chain sequences within a crystallite are in regular, highly extended conformations. The straightening and aligning of portions of the network chains thus decrease the deformation in the remaining amorphous regions, with an accompanying decrease in the stress.

In summary, the anomalous upturn in modulus observed for crystallizable polymers such as natural rubber and cis-1,4-polybutadiene is largely if not entirely due to strain-induced crystallization. In the case of the non-crystallizable PDMS model networks it is clearly due to limited chain extensibility, and thus the

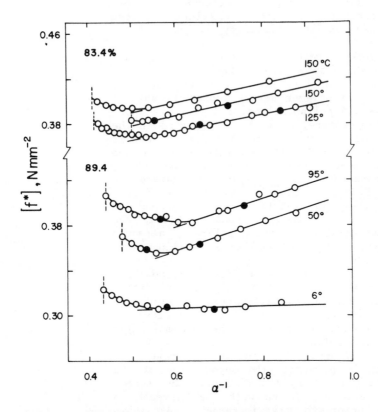

Figure 25. Typical results showing the effect of temperature on stress-strain isotherms obtained for bimodal PDMS networks (24). The curves are labelled with the mol% of short chains in the networks and the temperatures at which the elasticity measurements were carried out. The filled symbols represent checks for elastic reversibility, and the vertical lines locate the rupture points. (Reproduced with permission from Ref. 24. Copyright 1980, American Institute of Physics.)

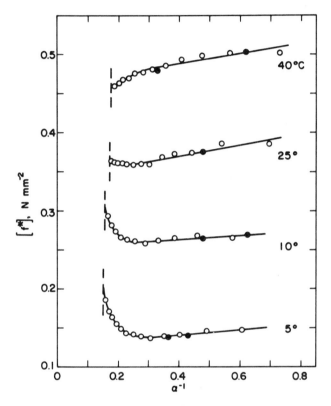

Figure 26. Stress-strain isotherms for highly crystallizable
cis-1,4-polybutadiene networks (25, 26). The curves have been
arbitrarily shifted along the ordinate to prevent overlapping
of some of the data points.

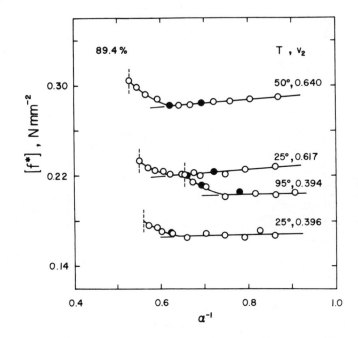

Figure 27. Typical results showing the effect of network swelling on stress-strain isotherms obtained for bimodal PDMS networks. The networks had 89.4 mol% short chains, and each isotherm is labelled with the temperature and volume fraction of polymer in the swollen network. (Reproduced with permission from Ref. 24. Copyright 1980, American Institute of Physics.)

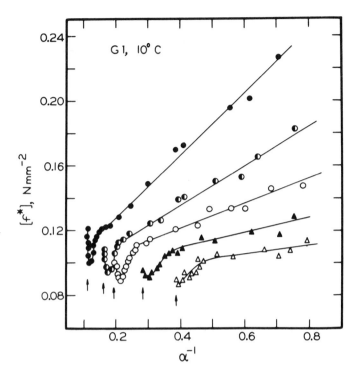

Figure 28. Stress-strain isotherms for a highly crystallizable cis-1,4-polybutadiene network swollen with 1,2-dichlorobenzene to values of the volume fraction v_2 of polymer of 1.00 (●), 0.80 (◖), 0.60 (○), 0.40 (▲), and 0.20 (△) (26, 27).

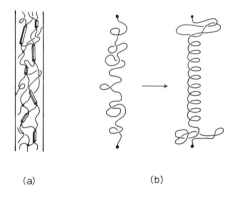

(a)   (b)

Figure 29. Strain-induced crystallization in a polymer network that has been elongated by a force along the vertical direction (9, 28).

results on this system will be extremely useful for the reliable evaluation of the various non-Gaussian theories of rubberlike elasticity.

ULTIMATE
PROPERTIES

This Section continues the discussion of unfilled elastomers at high elongations, but with an emphasis on ultimate properties, namely the ultimate strength and maximum extensibility.

Some illustrative results on the effects of strain-induced crystallization on ultimate properties are given for cis-1,4-polybutadiene networks in Table V (25). The higher the

Table V. Cis-1,4-Polybutadiene Networks
at High Elongation

T, °C	α at Upturn	Ultimate Properties	
		Max. Upturn in $[f^*]$, %	α at Rupture
5	3.27	54.2	6.64
10	3.48	30.1	6.22
25	4.03	4.3	5.85
40	--	0.0	5.68

temperature, the lower the extent of crystallization and, correspondingly, the lower the ultimate properties. The effects of increase in swelling parallel those for increase in temperature, since diluent also suppresses network crystallization. For non-crystallizable networks, however, neither change is very important, as is illustrated by the results shown for PDMS networks in Table VI (29). In the case of such non-crystallizable, unfilled elastomers, the mechanism for network rupture has been elucidated to a great extent by studies of model networks similar to those described in the preceding Section. For example, values of the modulus of bimodal networks formed by end-linking mixtures of very short and relatively long chains as illustrated in Figure 30 (15), were

Table VI. Ultimate Properties, PDMS Networks

v_2	λ_r^a	$\left[f^*\right]_r$
1.00	4.90	0.0362
0.80	4.42	0.0342
0.60	4.12	0.0338
0.40	4.16	0.0336

$^a\lambda_r$ is the value of the total elongation at the rupture point, using L_i(unswollen).

used to test the "weakest-link" theory, in which rupture was thought to be initiated by the shortest chains (because of their very limited extensibility). It was observed that increasing the number of very short chains did not significantly decrease the ultimate properties. The reason, given schematically in Figure 31 (24), is the very non-affine nature of the deformation at such high elongations. The network simply reapportions the increasing strain amongst the polymer chains until no further reapportioning is possible. It is generally only at this point that chain scission begins, leading to rupture of the elastomer. Some evidence in support of this conclusion is shown in Figure 32 (24). The weakest-link theory is implicitly based on the assumption of an affine deformation, which means that the elongation at which the modulus increases should be independent of the number of short chains in the network. The results in Figure 32 show the opposite behavior; the smaller the number of short chains, the easier the reapportioning and the higher the elongation required to bring about the upturn in modulus.

There turns out to be an exciting bonus if one puts a very large number of short chains into the bimodal network. The ultimate properties are then actually improved! This is illustrated in Figure 33 (30), in which data on PDMS networks are plotted in such a way that the area under a stress-strain

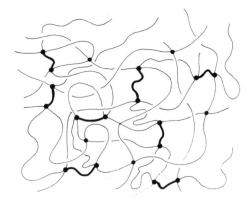

Figure 30. A portion of a network that is compositionally heterogeneous with respect to chain length. The very short and relatively long chains are arbitrarily shown by the thick and thin lines, respectively. (Reproduced with permission from Ref. 15. Copyright 1979, Hüthig & Wepf Verlag, Basel.)

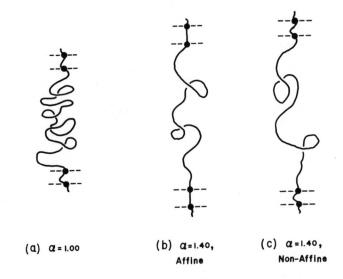

Figure 31. The effect of deformation on an idealized network segment consisting of a relatively long chain bracketed by two very short chains. (Reproduced with permission from Ref. 24. Copyright 1980, American Institute of Physics.)

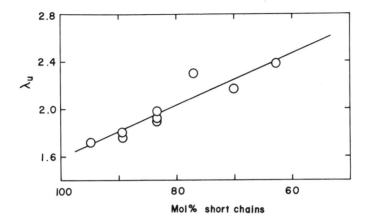

Figure 32. *The effect of PDMS network composition on the minimum elongation at which the upturn in modulus is discernible. (Reproduced with permission from Ref. 24. Copyright 1980, American Institute of Physics.)*

isotherm corresponds to the energy required to rupture the network. If the network is all short chains, it is brittle, which means that the maximum extensibility is very small. If the network is all relatively long chains, the ultimate strength is very low. In neither case is the material a tough elastomer. As can readily be seen from the figure, the bimodal networks are much improved elastomers in that they can have a high ultimate strength without the usual decrease in maximum extensibility.

A series of experiments were carried out in an attempt to determine if this reinforcing effect in bimodal PDMS networks could possibly be due to some intermolecular effect such as strain-induced crystallization. In the first experiment, illustrated in Figure 34 (31), temperature was found to have little effect on the isotherms, which strongly argues against the presence of any crystallization. So also do the results of stress-temperature and birefringence-temperature measurements (31). In a final experiment, the short chains were pre-reacted in a two-step preparative technique so as possibly to segregate them in the network structure, as shown in Figure 35 (32). This had very little effect on elastomeric properties, again arguing against any type of intermolecular organization as the origin for the reinforcing effects. Apparently, the observed increases in modulus are due to the limited chain extensibility of the short chains, with the long chains serving to retard the rupture process.

The molecular origin of the unusual properties of bimodal PDMS networks having been elucidated at least to some extent, it is now possible to utilize

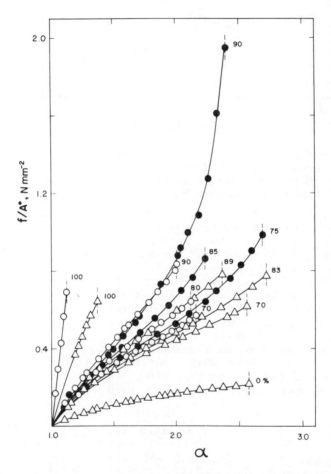

Figure 33. Typical plots of nominal stress against elongation for (unswollen) bimodal PDMS networks consisting of relatively long chains [M_c = 18,500 g mol^{-1}] and very short chains [1100 ($\triangle$), 660 ($\bigcirc$), and 220 ($\bullet$)]. Each curve is labelled with the mol% of short chains it contains, and the area under each curve represents the rupture energy (a measure of the "toughness" of the elastomer). (Reproduced with permission from Ref. 30. Copyright 1981, John Wiley & Sons, Inc.)

these materials in a variety of applications. The first involves the interpretation of limited chain extensibility in terms of the configurational characteristics of the PDMS chains making up the network structure ($\underline{14}$).

The first important characteristic of limited chain extensibility is the elongation α_u at which the increase in modulus first becomes discernible. Although the deformation is non-affine in the vicinity of the upturn, it is possible to provide at least a semi-quantitative interpretation of such results in terms of the network chain dimensions ($\underline{14,24}$). At the beginning of the upturn, the average extension r of a network chain having its end-to-end vector along the direction of stretching is simply the product of the unperturbed dimension $<r^2>_0^{\frac{1}{2}}$ and α_u ($\underline{24}$). Similarly, the maximum extensibility r_m is the product of the number n of

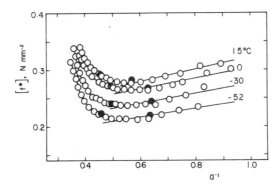

Figure 34. *Typical results obtained to determine the effect of temperature on the bimodal PDMS stress-strain isotherms; they pertain to networks containing 75 mol% of very short chains. (Reproduced with permission from Ref. 31. Copyright 1982, John Wiley & Sons, Inc.)*

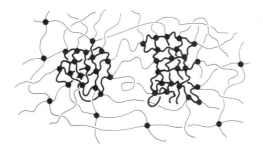

Figure 35. *Schematic sketch of a portion of a network that is spatially as well as compositionally heterogeneous with regard to chain length. (Reproduced with permission from Ref. 32. Copyright 1981,* Rubber Chem. Technol.)

skeletal bonds and the factor 1.34 Å which gives
the axial component of a skeletal bond in the most
extended helical form of PDMS, as obtained from the
geometric analysis summarized for PDMS and
polyethylene in Figure 36 (24). The ratio r/r_m at
α_u thus represents the fraction of the maximum
extensibility occurring at this point in the
deformation. The values obtained indicate that the
upturn in modulus generally begins at approximately
60-70% of maximum chain extensibility (24). This
is approximately twice the value which had been
estimated previously (3), in a misinterpretation of
stress-strain isotherms of elastomers undergoing
strain-induced crystallization.

It is also of interest to compare the values
of r/r_m at the beginning of the upturn with some
theoretical results by Flory and Chang (33,34) on
distribution functions for PDMS chains of finite
length. Of relevance here are the calculated values
of r/r_m at which the Gaussian distribution function
starts to over-estimate the probability of extended
configurations, as judged by comparisons with the
results of Monte Carlo simulations. The theoretical
results most relevant to the experimental results
on the bimodal PDMS networks are shown in Figure 37
(33). They suggest, for example, that the network
of PDMS chains having n = 53 skeletal bonds which
was experimentally studied should show an upturn at
a value of r/r_m a little less than 0.80. The
observed value was 0.77 (24), which is thus in
excellent agreement with theory.

A second important characteristic is the value
α_r of the elongation at which rupture occurs. The
corresponding values of r/r_m show that rupture
generally occurred at approximately 80-90% of
maximum chain extensibility (24). These quantitative results on chain dimensions are very important
but may not apply directly to other networks, in
which the chains could have very different
configurational characteristics and in which the
chain length distribution would presumably be quite
different from the very unusual bimodal
distribution intentionally produced in the present
networks.

Since dangling chains represent imperfections
in a network structure, one would expect their
presence to have a detrimental effect on the
ultimate properties, α_r and $(f/A^*)_r$, of an elastomer. This expectation is confirmed by an extensive
series of results obtained on PDMS networks which
had been tetrafunctionally cross-linked using a
variety of techniques. Some pertinent results are

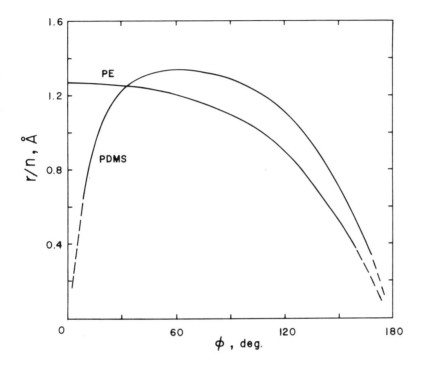

Figure 36. The end-to-end distance per skeletal bond n for regular conformations of polydimethylsiloxane and polyethylene network chains. The maximum extensibility r_m of the PDMS chain molecule occurs at $r_m/n = 1.34$ Å. (Reproduced with permission from Ref. 24. Copyright 1980, American Institute of Physics.)

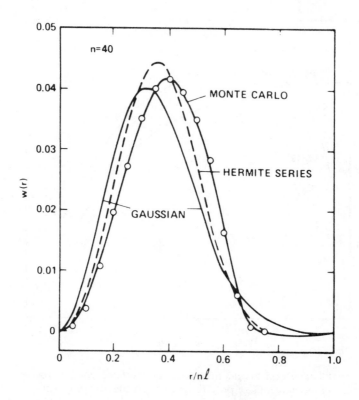

Figure 37. Radial distributions (in Å⁻¹) for PDMS chains having n = 40 skeletal bonds (each of length l = 1.64 Å) (33, 34).

shown, as a function of the molecular weight between cross-links, in Figure 38 (35,36). The largest values of $(f/A^*)_r$ are obtained for the networks prepared by selectively joining functional groups occurring either as chain ends or as side groups along the chains. This is to be expected, because of the relatively low incidence of dangling ends in such networks. (As already described, the effects are particularly pronounced when such model networks are prepared from a mixture of relatively long and very short chains). Also as expected, the lowest values of the ultimate properties generally occur for the networks cured by radiation (UV light, high-energy electrons, and γ radiation) (35). The peroxide-cured networks are generally intermediate to these two extremes, with the ultimate properties presumably depending on whether or not the free-radicals generated by the peroxide are sufficiently reactive to cause some chain scission. Similar results were obtained for the maximum extensibility (35). These observations are at least semi-quantitative and certainly interesting, but are somewhat deficient in that information on the number of dangling ends in these networks is generally not available.

More definitive results have been obtained by investigation of a series of model networks prepared by end-linking vinyl-terminated PDMS chains (35). The tetrafunctional end-linking agent was used in varying amounts smaller than that corresponding to a stoichiometric balance between its active hydrogen atoms and the chains' terminal vinyl groups. The ultimate properties of these networks, with known numbers of dangling ends, were then compared with those obtained on networks previously prepared so as to have negligible numbers of these irregularities (35).

Values of the ultimate strength of the networks are shown as a function of the high deformation modulus $2C_1$ in Figure 39 (35). The networks containing the dangling ends have lower values of $(f/A^*)_r$, with the largest differences occurring at high proportions of dangling ends (low $2C_1$), as expected. These results thus confirm the less definitive results shown in Figure 38. The values of the maximum extensibility show a similar dependence, as expected.

Another type of elastic deformation very widely used to characterize network structures is swelling (2,3,37). This is a three-dimensional dilation in

SWELLING AND OTHER TYPES OF DEFORMATION

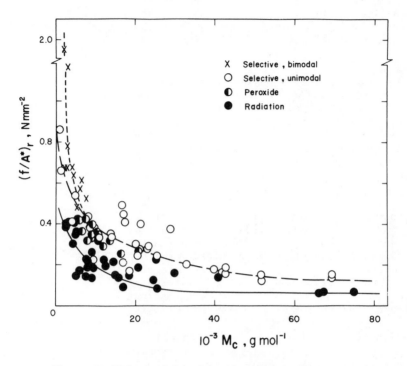

Figure 38. Values of the ultimate strength shown as a function of the molecular weight M_c between cross-links for (unfilled) tetrafunctional PDMS networks at 25 °C (36). (Reproduced with permission from Ref. 35. Copyright 1981, John Wiley & Sons, Inc.)

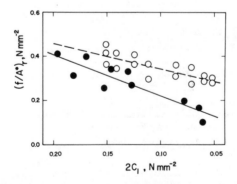

Figure 39. The ultimate strength shown as a function of the high deformation modulus for tetrafunctional PDMS networks containing a negligible number of dangling ends (○) and dangling ends introduced by using less than the stoichiometrically required amount of end-linking agent (●) (36). In the latter case, decrease in $2C_1$ corresponds to increase in the number of dangling ends. (Reproduced with permission from Ref. 35. Copyright 1981, John Wiley & Sons, Inc.)

which the network absorbs solvent, reaching an equilibrium degree of swelling at which the free energy decrease due to the mixing of the solvent with the network chains is balanced by the free energy increase accompanying the stretching of the chains. The classic swelling theory developed by Flory and Rehner gives the relationship ($\underline{2}$)

$$\nu/V = -\left[\ell n(1 - v_{2m}) + v_{2m} + \chi_1 v_{2m}^2\right]/$$

$$A_\phi' V_1 v_{2S}^{2/3} (v_{2m}^{1/3} - \omega v_{2m})\bigg] \qquad (25)$$

where ν/V is cross-link density, v_{2m} volume fraction of polymer at swelling equilibrium, χ_1 the usual free energy of interaction parameter ($\underline{2}$), A_ϕ' = 1 in the affine limit, V_1 is the molar volume of the solvent, v_{2S} the volume fraction of polymer present during cross-linking, and $\omega = 2/\phi$. If thermodynamic information ($\underline{2}$) on the strength of the polymer-solvent interactions is available, this method provides another valuable technique for determining the degree of cross-linking of a network.

In a refined theory developed by Flory ($\underline{37}$), the extent to which the swelling deformation is non-affine depends on the looseness with which the cross-links are embedded in the network structure. This depends in turn on both the structure of the network and its degree of equilibrium swelling. The resulting equation is

$$\nu/V = -\left[\ell n(1 - v_{2m}) + v_{2m} + \chi_1 v_{2m}^2\right]/$$

$$F_\phi V_1 v_{2S}^{2/3} v_{2m}^{1/3} \bigg] \qquad (26)$$

The factor F_ϕ characterizes the extent to which the deformation in swelling approaches the affine limit, and is given by

$$F_\phi = (1 - 2/\phi)\left[1 + (\mu/\xi)K\right] \qquad (27)$$

where ξ is the cycle rank of the network (the number of cuts required to eliminate all cyclic paths), $K = f(v_{2m}, \kappa, p)$ ($\underline{37}$), κ is a parameter specifying constraints on cross-links, and p a parameter specifying dependence of cross-link fluctuations on the strain ($\underline{37}$). This theory is somewhat more difficult to apply since it contains parameters not present in the simpler theory.

There are numerous other deformations of interest, including compression, biaxial extension, shear, and torsion. The equation of state for compression ($\alpha < 1$) is the same as that for elongation ($\alpha > 1$), and the equations for the other deformations may all be derived from Equation (10) by proper specification of the deformation ratios (2,3). Some of these deformations are considerably more difficult to study than simple elongation and, unfortunately, have therefore not been as extensively investigated.

SOME COMMENTS ON BIOELASTOMERS

There are a number of cross-linked proteins that are elastomeric, and their investigation may be used to obtain insights into elastic behavior in general. For example, elastin (38-41), which occurs in mammals, illustrates the relevance of several molecular characteristics to the achievement of rubberlike properties. Some relevant information is given in Figure 40. First, a high degree of chain flexibility is achieved in elastin by its chemically irregular structure, and by choices of side groups that are almost invariably very small. Since strong intermolecular interactions are generally not conducive to good elastomeric properties, the choices of side chains are also almost always restricted to non-polar groups. Finally, elastin has a glass transition temperature of approximately $200°C$ in the dry state, which means it would be elastomeric only above this temperature. Nature, however, apparently also knows about "plasticizers". Elastin, as used in the body, is invariably swollen with sufficient biological fluids to bring its glass transition temperature below the operating temperature of the body.

Another bioelastomer, found in some insects, is resilin (42). It is an unusual material because of its very high efficiency in storing elastic energy (i.e., very small losses due to viscous effects). A molecular understanding of this very attractive property could obviously have considerable practical as well as fundamental importance.

CURRENT PROBLEMS AND NEW DIREC-TIONS

Table VII lists some aspects of rubberlike elasticity which are in need of additional research. An example of a new cross-linking technique

Table VII.   Current Problems and New Directions

New cross-linking techniques

Improved understanding of network topology

More experimental results for deformations other than elongation

Generalization of phenomonological theory

Improved understanding of dependence of T_g and T_m on polymer structure

Preparation and characterization of "high-performance" elastomers

Study of possibly unique properties of bio-elastomers

Improved understanding of reinforcing effect of filler particles in a network

---

currently under development is the preparation of triblock copolymers such as those of styrene-butadiene-styrene. This system phase-separates in such a way that relatively hard polystyrene domains act as temporary, physical cross-links, as is shown in Figure 41 (43). The resulting elastomer is thermoplastic, and it is possible to reprocess it by simply heating it to above the glass transition temperature of polystyrene. It is thus a reprocessible elastomer.

High-performance elastomers are those which remain elastomeric to very low temperatures, and are relatively stable at very high temperatures. Some phosphazene polymers, shown schematically in Figure 42 (44,45), are in this category. These polymers have rather low glass transition temperatures, which is surprising since the skeletal bonds of the chains are thought to have some double-bond character. There are thus a number of interesting problems related to the elastomeric behavior of these unusual semi-inorganic polymers.

A particularly challenging problem is the development of a more quantitative molecular understanding of the effects of filler particles, such as carbon black in natural rubber and silica

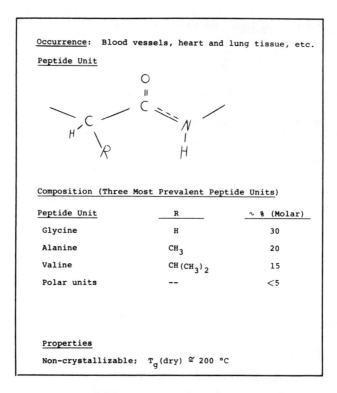

Occurrence: Blood vessels, heart and lung tissue, etc.

Peptide Unit

Composition (Three Most Prevalent Peptide Units)

Peptide Unit	R	~ % (Molar)
Glycine	H	30
Alanine	CH_3	20
Valine	$CH(CH_3)_2$	15
Polar units	--	<5

Properties

Non-crystallizable; T_g(dry) $\cong$ 200 °C

Figure 40. Some structure-property information on the elasto-meric protein elastin (38-41).

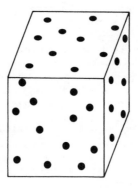

Figure 41. Sketch of a multiphase, thermoplastic elastomer.

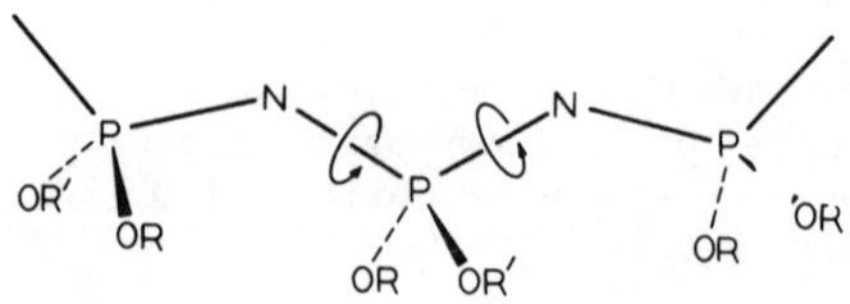

Figure 42. Sketch of a semi-inorganic phosphazene polymer (45). (Reproduced with permission from Ref. 44. Copyright 1977, John Wiley & Sons, Inc.)

in silicone polymers (46-49). Such fillers provide tremendous reinforcement in elastomers, and how they do this is still only poorly comprehended. It is difficult to imagine another problem of comparable practical significance in the area of rubberlike elasticity.

A.   Analysis of Some Typical Elongation or Compression Data

Suppose a network having tetrafunctional crosslinks ($\phi = 4$, $A_\phi = 1/2$) and a density of 0.900 g cm^{-3} has $[f^*]$ ($\alpha = \infty$) = 0.100 N mm^{-2} $[10^5$ N m^{-2}(Pa) = 10^{-1} MN m^{-2}(MPa) = 1.02 kg cm$^{-2}]$ at 298.2K. Calculate the network chain density, the cross-link density, and the average molecular weight between cross-links.

B.   Analysis of Some Typical Swelling Data

A typical network studied in this regard might have been tetrafunctionally crosslinked in the undiluted state ($v_{2S} = 1.00$), and exhibit an equilibrium degree of swelling characterized by v_{2m} = 0.100 in a solvent having a molar volume V_1 = 80 cm^3 mol^{-1} (8.00 x 10^4 mm^3 mol^{-1}) and an interaction parameter with the polymer corresponding to X_1 = 0.30. Calculate the network chain density.

A.   Use of the above data in Equation (17) with k in units (1.381 x 10^{-20} N mm K^{-1} chain^{-1}) compatible with $[f^*](\alpha = \infty)$ in N mm^{-2} gives

$$\nu/V = 4.86 \times 10^{16} \text{ chains mm}^{-3}$$

Use of Avogadro's number N_{Avo} = 6.02 x 10^{23} mol^{-1}, then gives

$$\nu/V = 8.06 \times 10^{-8} \text{ mols of chains mm}^{-3}$$

As specified by Equation (23), the density of crosslinks would be one-half (2/ϕ) of this value,

$$\nu/V = 4.03 \times 10^{-8} \text{ mols of crosslinks mm}^{-3}$$

Since the polymer has a density $\rho = 0.900$ g cm^{-3} (9.00×10^{-4} g mm^{-3}), Equation (24) indicates

$$M_c = 1.12 \times 10^4 \text{ g mol}^{-1}$$

B. The standard swelling relationship [Equation (25)] with $A_\phi' = 1$ would give

$$\nu/V = 7.13 \times 10^{-8} \text{ mols of chains mm}^{-3}$$

Use of the improved relationship [Equation (26)] with the reasonable estimates (37) $\kappa = 20$ and $p = 2$ gives $K = 0.42$ (37) and thus

$$\nu/V = 8.95 \times 10^{-8} \text{ mols of chains mm}^3$$

This result is seen to be not very different from the value calculated using the simpler relationship given in Equation (25).

LITERATURE
CITED

1. J. E. Mark, J. Chem. Educ., **58**, 898 (1981).
2. P. J. Flory, "Principles of Polymer Chemistry", Cornell University Press, Ithaca, N.Y., 1953.
3. L. R. G. Treloar, "The Physics of Rubber Elasticity", 3rd ed., Clarendon, Oxford, 1975.
4. G. Ronca and G. Allegra, J. Chem. Phys., **63**, 4990 (1975).
5. P. J. Flory, Proc. R. Soc. London, Ser. A, **351**, 351 (1976).
6. P. J. Flory, Polymer, **20**, 1317 (1979).
7. P. J. Flory and B. Erman, Macromolecules, **15**, 800 (1982).
8. J. E. Mark, Rubber Chem. Technol., **48**, 495 (1975).
9. J. E. Mark, Acc. Chem. Res., **12**, 49 (1979).
10. P. J. Flory, A. Ciferri, and C. A. J. Hoeve, J. Polym. Sci., **45**, 235 (1960).
11. A. Ciferri, C. A. J. Hoeve, and P. J. Flory, J. Am. Chem. Soc., **83**, 1015 (1961).
12. J. E. Mark, Rubber Chem. Technol., **46**, 593 (1973).
13. J. E. Mark, Macromolecular Rev., **11**, 135 (1976).
14. P. J. Flory, "Statistical Mechanics of Chain Molecules", Wiley-Interscience, New York, 1969.
15. J. E. Mark, Makromol. Chemie, Supp. **2**, 87 (1979).

16. J. E. Mark, R. R. Rahalkar, and J. L. Sullivan, J. Chem. Phys., **70**, 1794 (1979).
17. M. A. Llorente and J. E. Mark, Macromolecules, **13**, 681 (1980).
18. J. E. Mark, Rubber Chem. Technol., **55**, 762 (1982).
19. N. R. Langley, R. A. Dickie, C. Wong, J. D. Ferry, R. Chasset, and P. Thirion, J. Polym. Sci., Part A-2, **6**, 1371 (1968).
20. R. M. Johnson and J. E. Mark, Macromolecules, **5**, 41 (1972).
21. C. U. Yu and J. E. Mark, Macromolecules, **7**, 229 (1974).
22. L. Mullins, J. Appl. Polym. Sci., **2**, 257 (1959).
23. J. E. Mark, M. Kato, and J. H. Ko, J. Polym. Sci., Part C, **54**, 217 (1976).
24. A. L. Andrady, M. A. Llorente, and J. E. Mark, J. Chem. Phys., **72**, 2282 (1980).
25. T.-K. Su and J. E. Mark, Macromolecules, **10**, 120 (1977).
26. J. E. Mark, Polym. Eng. Sci., **19**, 254 (1979).
27. D. S. Chiu, T.-K. Su, and J. E. Mark, Macromolecules, **10**, 1110 (1977).
28. J. E. Mark, Polym. Eng. Sci., **19**, 409 (1979).
29. D. S. Chiu and J. E. Mark, Colloid and Polymer Science, **254**, 644 (1977).
30. M. A. Llorente, A. L. Andrady, and J. E. Mark, J. Polym. Sci., Polym. Phys. Ed., **19**, 621 (1981).
31. Z.-M. Zhang and J. E. Mark, J. Polym. Sci., Polym. Phys. Ed., **20**, 473 (1982).
32. J. E. Mark and A. L. Andrady, Rubber Chem. Technol., **54**, 366 (1981).
33. P. J. Flory and V. W. C. Chang, Macromolecules, **9**, 33 (1976).
34. J. E. Mark in "Elastomers and Rubber Elasticity", ed. by J. E. Mark and J. Lal, American Chemical Society, Washington, D.C., 1982.
35. A. L. Andrady, M. A. Llorente, M. A. Sharaf, R. R. Rahalkar, J. E. Mark, J. L. Sullivan, C. U. Yu, and J. R. Falender, J. Appl. Polym. Sci., **26**, 1829 (1981).
36. J. E. Mark, Adv. Polym. Sci., **44**, 1 (1982).
37. P. J. Flory, Macromolecules, **12**, 119 (1979).
38. "Elastin and Elastic Tissue", ed by L. B. Sandberg, W. R. Gray, and C. Franzblau, Plenum, New York, 1977.
39. R. Ross and P. Bornstein, Sci. Am., **224**, 44 (1971).

40. C. A. Hoeve and P. J. Flory, Biopolymers, 13, 677 (1974).

41. A. L. Andrady and J. E. Mark, Biopolymers, 19, 849 (1980).

42. M. Jensen and T. Weis-Fogh, Phil. Trans. R. Soc. London, Ser. B., 245, 137 (1962).

43. S. L. Aggarwal, Polymer, 17, 938 (1976).

44. J. E. Mark and C. U. Yu, J. Polym. Sci., Polym. Phys. Ed., 15, 371 (1977).

45. A. L. Andrady and J. E. Mark, Eur. Polym. J., 17, 323 (1981), and pertinent references cited therein.

46. G. Kraus, Rubber Chem. Technol., 51, 297 (1978).

47. B. B. Boonstra, Polymer, 20, 691 (1979).

48. Z. Rigbi, Adv. Polym. Sci., 36, 21 (1980).

49. K. E. Polmanteer and C. W. Lentz, Rubber Chem. Technol., 48, 795 (1975).

SELECTED GENERAL BIBLIOGRAPHY

(A) P. J. Flory, "Principles of Polymer Chemistry", Cornell University Press, Ithaca, N.Y., 1953.

(B) P. Meares, "Polymers: Structure and Bulk Properties", Van Nostrand, New York, 1965.

(C) K. Dušek and W. Prins, Advan. Polym. Sci., 6, 1 (1969).

(D) "Polymer Networks: Structure and Mechanical Properties", ed. by A. J. Chompff and S. Newman, Plenum, New York, 1971.

(E) K. J. Smith, Jr., in "Polymer Science", ed. by A. D. Jenkins, North-Holland, Amsterdam, 1972.

(F) "Rubber Technology", ed. by M. Morton, Van Nostrand-Reinhold, New York, 1973.

(G) F. T. Wall, "Chemical Thermodynamics", Third Ed., Freeman, San Francisco, 1974.

(H) "Rubber and Rubber Elasticity", ed. by A. S. Dunn, Polymer Symposium 48, Wiley-Interscience, New York, 1974.

(I) L. R. G. Treloar, "The Physics of Rubber Elasticity", Third Ed., Clarendon Press, Oxford, 1975.

(J) "Chemistry and Properties of Crosslinked Polymers", ed. by S. S. Labana, Academic Press, New York, 1977.

(K) "Science and Technology of Rubber", ed. by F. R. Eirich, Academic Press, New York, 1978.

(L) L. K. Nash, J. Chem. Educ., 56, 363 (1979).

(M) J. E. Mark, J. Chem. Educ., 58, 898 (1981).

(N) "Elastomers and Rubber Elasticity", ed. by J. E. Mark and J. Lal, ACS, Washington, 1982.

The Glassy State and the Glass Transition

Adi Eisenberg

The glass transition is perhaps the most important single parameter which one needs to know before one can decide on the application of the many non-crystalline polymers that are now available. For example, if the glass transition temperature of a polymeric material is approximately -70°C, one would not consider that particular polymer for a window application, but rather for a rubber tire or tubing application. Conversely, if the glass transition temperature is +100°C, utilization as a rubber will be completely excluded, but any of the hard plastic applications will be considered.

In structural terms, glasses are characterized by the absence of long range order. Order along a polymer chain in the polymeric glasses is, naturally, to be expected, as well as some short range liquid-like order which is encountered not only in glasses but also in regular liquids. Many polymers are partly crystalline, and the crystalline regions are subject to the same behaviour as crystals in general.

It is the noncrystalline regions, however, that exhibit the phenomena that will be the subject of the present discussion. It is worth noting that it is not only polymers that exhibit these phenomena, but also a very wide range of other materials, and that the phenomenology discussed here is relatively new. The phenomena accompanying vitrification were not really explored until the 20's, and a complete theoretical understanding of the glass transition temperature is still not available. This is so in spite of the fact that a very large number of original publications are available on the glass transition

0851/84/0055$11.25/1
© 1984 American Chemical Society

specifically, and on glass transition phenomena in
general, in addition to several reviews (1-10).

In this chapter, the first part of the discussion
will be devoted to a presentation of the phenomenology
of the glass transition, stressing particularly
volumetric and mechanical properties. A brief introduction into the theories of the glass transition will
follow, stressing a very straightforward free volume
theory, but mentioning also the thermodynamic theories
as well as a kinetic approach. The next section will
be devoted to a discussion of the molecular parameters
and their effect on the glass transition. Parameters
such as chain stiffness, internal plasticization, and
intramolecular forces will be included. From another
point of view, the question will be asked as to why
polystyrene has a very different glass transition
temperature from poly-n-butyl acrylate on the one
hand and poly(dimethyl siloxane) on the other. The
subsequent section will be devoted to an exploration
of controllable parameters and their effect on the
glass transition. A discussion will be presented of
the factors that can change the glass transition
temperature of a particular polymer. More specifically, once a material like polystyrene has been chosen
as the substance to be explored, we will ask ourselves
what can affect its glass transition temperature.
Factors such as pressure, the presence of diluent, the
molecular weight, as well as the presence of comono-
mers, and various other parameters will be explored.
The next section will deal with methods of determining
the glass transition temperature. A very large number
of these are available, and only some of the most
important ones will be described briefly to introduce
the most common techniques. The chapter will conclude
with a discussion of molecular motions in the glassy
state and methods of measuring them. Specifically,
methods will be described which allow one to determine
which piece of a molecule in a glassy material moves
at what frequency as a function of temperature. The
experimental techniques here are quite well established; however, the results of the experiments, i.e.,
the molecular aspects, are still in the process of
exploration, and, as such, represent a most challenging and interesting field. This area also has many
practical implications, in that the presence of these
low temperature molecular motions, in many cases,
renders the material ductile as opposed to brittle,
and improves a wide range of other physical properties.

Before concluding the introduction, let us look
into the types of materials that exhibit the phenomena
which will be discussed. Naturally, most of the stress
will be placed on the polymeric materials, the so-
called linear synthetic high polymers. However, a
very wide range of other materials show exactly the

same phenomena in the vicinity of the glass transition. The vitrification process is exceedingly similar for this entire range of materials, which is summarized in Table I.

Table I.   Materials Exhibiting Glass Transition Phenomena (4).

Network Glasses	$(SiO_2, B_2O_3, P_2O_5, As_2S_3...)$
Modified Networks	$(SiO_2 + Na_2O, P_2O_5 + K_2O...)$
Linear or Branched Polymers	(PS, PE,...)
H - bonded Glasses	(Glycerine...)
Salts or Salt Mixtures	$(ZnCl_2 \ BeF_2, K_2CO_3 \cdot MgCO_3...)$
Electrolyte Solutions	$(H_2O - HCl \ Eutectic...)$
Metals	$(Pd_{80}Si_{20}, Fe_{40}Ni_{40}P_{14}B_6...)$
Low M.W. Materials	(2- methyl pentane...)

The list starts with the network glasses, i.e., materials like SiO_2, B_2O_3, P_2O_5, etc... These are three-dimensional polymers in which the repeat units are tri- or tetrafunctional moieties like those found in P_2O_5 or SiO_2. If one introduces into these networks network modifiers such as Na_2O or K_2O, some of the network points get ruptured, and an Si - O - Si bridge, for example, is converted to two - $SiO^{\ominus} \ Na^{\oplus}$ groups. Some linear segments are to be expected in these modified networks; for example, in the case of an equimolar mixture of Na_2O and P_2O_5, a predominantly linear polymer, $(NaPO_3)_x$, is found. In contrast to the network glasses based on silica, for example SiO_2, in which the glass transition temperature is of the order of 1200°C, the modified networks based on silica (for example, the soda - lime - silica glasses), have a glass transition temperatures in the vicinity of 500°C, a considerable decrease.

The next category includes the large number of linear or branched polymers which will be the main subject of this presentation. Here are included all the organic polymers that are normally available commercially, as well as some of the inorganics, for example, those based on sulfur or selenium, or even the phosphates. Under other classification schemes these might be included under the modified networks. Numerically, this is perhaps the largest category of all the materials listed in Table I.

The next group contains the hydrogen bonded systems, for example glycerine. Here, the glass transition temperatures are still lower, with glycerine, for example, having a T_g of approximately -80°C. The next category includes salts or salt mixtures of relatively low molecular weight; examples of materials within this group would be zinc chloride, beryllium chloride, potassium carbonate-magnesium, carbonate mixtures, or potassium nitrate - calcium nitrate mixtures. These have glass transition temperatures which are considerably lower than those of the modified networks, but usually somewhat higher than those of the normal organic polymers. Electrolyte solutions are still another category, and here one would encounter, for example, the water - HCl eutectic, which has a glass transition temperature well below -100°C.

Perhaps the most surprising category of all is the glassy metals. Here are included materials which are prepared from liquid metals, usually mixtures of several elements, by rapid cooling, involving cooling rates of the order of 100,000 to 1,000,000°/sec. These glassy metals have glass transition temperatures that can span a fairly wide region, the range of 0°C to +200° being typical.

The last category includes low molecular weight materials, such as 2-methyl pentane, mixtures of methyl tetrahydrofuran and toluene or various other non-linear hydrocarbons. These materials, when cooled at reasonably rapid rates, vitrify quite easily to give materials whose glass transition temperatures are of the order of -150 to -200°C.

An enormous range of materials is included in Table I. The most important of these, from the point of view of this discussion are the linear, lightly branched, or lightly crosslinked synthetic organic high polymers.

PHENOMENOLOGY OF THE GLASS TRANSITION (2-10)

The discussion of the phenomenology of the glass transition begins most conveniently with a study of the volumetric properties of polymeric materials, especially those which can either crystallize or vitrify, depending on the method of handling.

Let us start with point A in Figure 1(a), which gives a plot of the volume as a function of temperature of a crystalline material. As this material is heated, the volume expands in a perfectly normal fashion until one approaches the melting point. Here, if the material is completely crystalline and possesses a sharp melting point, the volume changes discontinuously and increases appreciably over very small temperature range. For normal polymeric materials, on the other hand, melting is a process which takes place over a fairly wide temperature range. For this reason, the volume increases more in a manner indicated by the solid line rather than discontinuously, as is observed in materials of low molecular weight. At the melting point, the material has lost all traces of crystallinity; it no longer shows the sharp X-ray bands which are expected from partly or completely crystalline materials, but shows only an amorphous halo. As one heats the material beyond the melting point, in the region of the graph between points C and D one observes a much more rapid expansion than that observed below the melting point, suggesting that here the packing is much less orderly than in the crystal, as one would expect, and much more free volume is introduced as the material is heated. This concept of free volume, while not very quantitative, is extremely useful and will be referred to repeatedly in the subsequent discussion. Far above the melting point, e.g., at point D, the material is a liquid of a reasonably high viscosity, depending on the molecular weight.

As one now starts cooling, one can cool most polymeric materials beyond point C without crystallization. Some polymers cannot be prevented from crystallizing (polyethylene is one of them), but many others, for instance isotactic polystyrene, can be cooled below the crystalline melting point without the slightest danger of crystallization; as a matter of fact, in some cases, it is extremely difficult or even impossible to crystallize polymers from the melt. As one keeps cooling beyond point C, one cannot continue indefinitely while still maintaining a more or less linear relationship between the volume and the temperature. Within a relatively narrow temperature region, indicated by the letter E in Figure 1, the volume – temperature plot changes slope dramatically, and below that temperature range, the expansion coefficient, which is the slope of this plot, becomes very much smaller than it was for the liquid. The glass transition temperature has been defined as the intersection of the straight line segments of the volume – temperature plot in the vicinity of point E. Experimentally, however, the glass transition is not a precise point, but a relatively narrow temperature region. In volume-temperature plots, as in this particular case, it is defined by a discontinuity in the rate of change of volume with respect to temperature, that is, by a change in the expansion coefficient. This will be described somewhat more extensively below.

A very similar range of phenomena is observed if one plots the enthalpy as a function of temperature. This is shown in Figure 1(b), the phenomenology being similar to that shown in Figure 1(a) for the volume.

A good idea of the kinetics of the vitrification process can be obtained if one explores the phenomena in the vicinity of point E while cooling the polymer at different cooling rates (4). One starts out in each case at an equilibrium volume far above the glass transition temperature, and cools in the first experiment rapidly until the change in slope is

observed in the volume-temperature plots in Figure 2.
The inflection point, or the glass transition temperature, is observed at a reasonably high value, i.e.,
point A. This is in marked contrast to what happens
when one cools at a very slow rate, which yields the
inflection point at a considerably lower temperature,
point B. Intermediate cooling rates would lead to
an inflection point at intermediate temperatures.
This type of experiment shows clearly that the glass
transition temperature, as determined in these
straightforward cooling experiments, is a function
of the cooling rate. A ballpark figure of the effect
of cooling rate on the glass transition temperature
is approximately a 3% change per order of magnitude
change in the cooling rate. Thus, a material which
has been cooled at 100° per minute, will have a glass
transition temperature approximately 6° higher than
the same material cooled at 1° per minute. It is
worth keeping this in mind when considering the glass
transition temperature of a rapidly quenched sample,
which will be quite different from that of a slowly
cooled sample.

Another interesting aspect of the glass transition can be seen when one cools the material very
rapidly to point A in Figure 2, keeps it at a constant temperature, and monitors the volume as a function of time at this constant temperature. Depending
on the detailed cooling rate and the temperature to
which one has cooled the sample, behaviour of the
type found in Figure 3 will be encountered (4). In
that figure, the volume at any time T, minus the
equilibrium volume, in other words, the volume that
is attained on storage of the sample for a very long
time, is plotted as a function of the logarithmic
time; several different curves are seen depending on
the temperature to which a sample has been quenched.
The T_g in Figure 3 is considered to be the "normal"
T_g, that is, the point obtained at a cooling rate of
approximately 1° per minute. Thus, one finds that
if a material is cooled very rapidly to a temperature
T_g-10 and is stored at that temperature for some
time, the volume shrinks along the line shown by the
curve for T_g-10. At T_g-5 the shrinkage will be somewhat more rapid, and still more rapid at T_g or at
T_g+2. In any case, the volume of a material which
has been quenched very rapidly to a particular
temperature in the vicinity of the glass transition
and maintained at that temperature will continue
shrinking. This is a most significant point and of
great commercial importance if the sample that is
produced is expected to retain its structural integrity and dimensional stability over a long period
of time.

Since enthalpy exhibits phenomena very similar

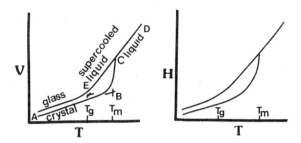

Figure 1. Volume-temperature and enthalpy-temperature relations for glass-forming materials.

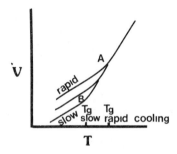

Figure 2. Volume-temperature behavior as a function of cooling rate.

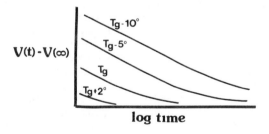

Figure 3. Isothermal volume contraction near the glass transition temperature. (Adapted from Ref. 4.)

to those of volume, it is not surprising that enthalpy relaxation also takes place and that thermodynamic properties of a polymer change just as its volume changes with storage in the vicinity of the glass transition temperature.

On occasion, when looking at plots of the volume as a function of temperature, unusual hysteresis effects are observed (11). Two examples are shown in Figure 4. Let us first consider what happens when the sample is cooled very rapidly along the line marked with the number 1 and subsequently heated, very much more slowly than the rate at which was cooled, along the dotted line marked 2. The sample, as it warms up, has much more time to adjust to its environment than it had while it was cooling. This longer time availability results in a volume shrinkage, with the result that the volume of the sample near T_g on heating is considerably lower than the volume that it had at corresponding temperatures on cooling. Storing the sample for long periods of time, as was shown in Figure 3, results in a volume shrinkage; a slow heating rate following a rapid cooling thus has the same effect.

The opposite effect is observed when one cools the sample very slowly and heats it rapidly along the lines 3 and 4 in the same figure. As expected, slow cooling results in a very low apparent glass transition temperature. However, a very rapid heating of a slowly cooled sample results in an overshooting of the original glass transition temperature, and leads to a very rapid expansion considerably above the inflection point of the cooling curve. This again leads to a serious hysteresis effect, which makes the determination of the glass transition temperature from volumetric properties alone somewhat uncertain if the heating and cooling rates are unequal. This factor points out the need of maintaining equal heating and cooling rates when determining glass transition temperatures by these techniques.

Finally, it is worth looking at the first derivatives of either the curve of volume versus temperature or the curve of enthalpy versus temperature. The first derivative of the volume with respect to temperature curve, divided by the volume, $\left(\frac{1}{V}\frac{\delta V}{\delta T}\right)_p$, is simply the expansion coefficient, α. Both v versus T and α versus T plots are shown in Figure 5(a), indicating clearly that there is a discontinuity in the first derivative of volume versus time, not just a change in slope. As a result of the presence of this discontinuity, the glass transition temperature, in much of the early literature, has been referred to as a second order transition. It should be stressed that the glass transition temperature, as it is observed, is not a

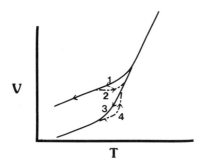

Figure 4. Hysteresis effects in volume–temperature plots near T_g. (Adapted from Ref. 11.)

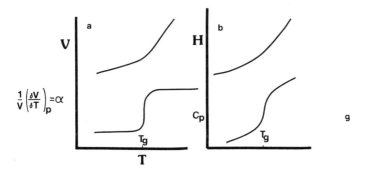

Figure 5. Temperature dependence of a, volume and expansion coefficient and b, enthalpy and heat capacity near T_g.

second order transition but a kinetic phenomenon which may very well have underlying thermodynamic reasons, as will be shown later. Similar curves of the first derivative of the enthalpy as a function of temperature are shown in Figure 5(b). The heat capacity, which is the first derivative of the enthalpy with respect to temperature at constant pressure, shows a similar discontinuity to that observed in the expansion coefficient, explaining the historical reference to the glass transition as a second order transition.

Having considered the volumetric, or more generally, thermodynamic properties as a function of temperature, we can now proceed to a brief consideration of the mechanical properties, specifically the viscosity, the modulus, and loss tangent or energy dissipation. Figure 6 shows a plot of the logarithm of the viscosity as a function of the temperature for polymeric materials (11). In the vicinity of the glass transition temperature, no dramatic discontinuities of the type observed in the plot of the volume as a function of temperature are seen; however, one characteristic feature for an extremely wide range of non-crystalline materials is the fact that the viscosity at the glass transition temperature is of the order of 10^{13} poise. This is true independent of whether the material is a linear organic polymer, a splat cooled metal, or one of the low molecular weight materials, organic or inorganic. It should be mentioned that the detailed shape of the plot at low temperatures is the subject of considerable controversy. The phenomenology near T_g, however, is well explored.

A similar absence of dramatic discontinuities is observed in a plot of the modulus as a function of temperature. For viscoelastic materials, the modulus is a function not only of the temperature of the measurement but also of the time. Therefore, it is necessary to fix a time scale of experiment, and in this particular case, one considers the ten second modulus. Thus, if one plots the ten second modulus as a function of temperature, as is shown in Figure 7, four distinct regions of behaviour are seen (11). At very low temperatures, the modulus has a value of the order of 10^{10} dynes/cm^2, typical of organic glasses. Over a fairly narrow temperature range between T_g-20 and T_g+20, the modulus drops by approximately three orders of magnitude, and sometimes even more, to a value of the order of $10^{6.5}$ dynes/cm^2, a value typical of lightly crosslinked rubbers or of physically entangled long chains. It can drop very rapidly beyond that point as the temperature increases, if the material is not crosslinked. However, in the presence of crosslinks, the material can exhibit a rubbery plateau at a value of the Young's modulus of $10^{6.5}$ dynes/cm^2 (or higher) which can extend over

considerable temperature ranges. If the material is
not crosslinked, the modulus drops further at a
temperature depending on the molecular weight. In
the glass transition region, for a very wide range of
materials, it has been found that the modulus has a
value of the order of 10^9 dynes/cm^2. This is not true
if the material contains very strong intermolecular
forces, for instance ionic groups, in which case the
modulus would have a much higher value at the glass
transition temperature. However, for normal organic
polymers such as polystyrene, poly(methyl methacrylate), or polybutadiene, a value of 10^9 dynes/cm^2 is
characteristic of the vicinity of the glass transition
temperature.

Finally, the behaviour of the loss tangent as a
function of temperature in the vicinity of the glass
transition will be discussed. The loss tangent is a
measure of the energy dissipation, and many experimental techniques are available for its determination.
For example, if a rubber ball were dropped from a
particular height and it bounced back to almost the
same height, the loss tangent of that material would
be fairly low; on the other hand, if it rebounded only
slightly, the loss tangent would be quite high. The
higher the loss tangent, or the greater the amount of
energy that is dissipated on impact, the lower the
bounce of the ball. Many quantitative methods of
determining the loss tangent are available, but these
are not of importance at this point.

When the energy dissipation of a polymer is
investigated as a function of temperature at constant
frequency, a plot of the type found in Figure 8 is
observed (13). A number of low temperature energy
dissipation peaks are seen, as well as a very dramatic
peak at the glass transition temperature. In many
cases, this loss tangent maximum is taken as the glass
transition temperature. A cautionary comment, however,
is called for. It should be noted that the position
of the peak in dynamic experiments is a function of
the frequency of the experiment, so that in comparing
results obtained from studies such as volume-
temperature measurements with those from dynamic
mechanical studies, comparable frequencies or comparable cooling rates must be involved. The volumetric
or thermodynamic results for a material studied at a
cooling rate of 1° per minute should be compared to
the loss tangent peak obtained at a frequency of
about one cycle per minute; comparisons over drastically different frequency or cooling rate ranges may lead
to erroneous results. The loss tangent will be
encountered again when molecular motions are discussed
in the last part of this presentation.

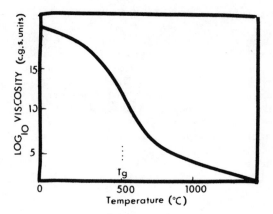

Figure 6. *Viscosity-temperature plot for a soda-lime-silica glass. (Reproduced with permission from Ref. 11. Copyright 1956, Methuen.)*

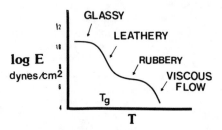

Figure 7. *Modulus-temperature plot for a typical linear polymer. (Adapted from Ref. 12.)*

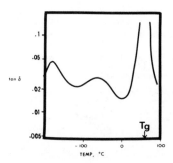

Figure 8. *Loss tangent vs. temperature at ca. 1 Hz for poly[2-methyl-6-(2-octyl)phenylene oxide]. (Adapted from Ref. 13.)*

As mentioned before, a complete theoretical understanding of the glass transition phenomenon is not yet available. The theories which have been proposed can be divided into three categories: free volume theories, thermodynamic theories, and kinetic theories. In the present discussion, only one free volume theory will be described briefly, to be followed by a brief mention of a thermodynamic theory as well as one kinetic approach.

The free volume theory to be discussed here starts with the Dolittle equation (16) which is given as Equation 1.

$$\eta = A \exp (B \, V_o/V_f) \tag{1}$$

Here η is the viscosity, V_o and V_f are the occupied and the free volumes respectively, and A and B are constants. This provides the theoretical background for the so-called Williams Landell and Ferry equation (commonly known as a W.L.F. equation), which is of enormous importance in the study of polymer viscoelasticity. Taking the logarithm of Equation 1, one obtains Equation 2.

$$\ln \eta = \ln A + B \, V_o/V_f \tag{2}$$

Defining $f = \dfrac{V_f}{V_o + V_f} \simeq \dfrac{V_f}{V_o}$ (since $V_o \gg V_f$)

one can rewrite Equation 2 as

$$\ln \eta = \ln A + B \, \frac{1}{f} \tag{3}$$

If we now let the free volume at the glass transition temperature be f_g, and if we allow it to increase above the glass transition temperature with the expansion coefficient α_f, which is approximately equal to the liquid state expansion coefficient minus the glassy state expansion coefficient (or to the difference in expansion coefficients above and below the glass transition temperature), we can describe the fractional free volume at any temperature above the glass transition, f_t, by Equation 4.

$$f_t = f_g + \alpha_f \, (T-T_g) \tag{4}$$

Let us now consider the ratio of the viscosities, i.e., η_T, the viscosity at any temperature T, divided by η_{T_g} the viscosity at T_g, utilizing Equation 3 along with equation 4 inserted for $1/f$. The results are given in Equation 5.

$$\ln \frac{\eta_T}{\eta_{T_g}} = \ln a_T = B(\frac{1}{f_T} - \frac{1}{f_g}) \tag{5}$$

THEORIES OF THE
GLASS TRANSITION
(5, 14, 15)

where a_T is simply η_T/η_{T_g}. Inserting Equation 4 for the fractional free volume at the temperature T into Equation 5, we obtain after appropriate manipulations, Equation 6.

$$\log a_T = B\left(\frac{1}{f_g + \alpha_f(T-T_g)} - \frac{1}{f_g}\right)$$

$$= \frac{B}{f_g}\left[\frac{f_g - f_g - \alpha_f(T-T_g)}{f_g + \alpha_f(T-T_g)}\right]$$

$$= -\frac{B}{f_g}\frac{T-T_g}{(f_g/\alpha_f) + T - T_g}$$

$$= -\frac{B}{2.3f_g}\frac{T-T_g}{(f_g/\alpha_f) + T - T_g} \tag{6}$$

It is worth noting that the constant B was found to be very close to unity. Equation 6 is a form of the W.L.F. or Williams Landell Ferry equation. That equation has been given in terms of the so-called universal parameters, as in Equation 7, which is the W.L.F. equation (17).

$$\log a_T = -17.4 \frac{T-T_g}{51.6 + T - T_g} \tag{7}$$

The constants 17.4 and 51.6 are nearly universal, being valid, within a reasonable degree of approximation, for a very wide range of materials. The value 17.4 for the first constant implies that $1/2.3f_g$ is equal to 17.4 or that the fractional free volume, f_g, at the glass transition temperature is approximately .025, as shown in Equation 8.

$$\frac{1}{2.3f_g} \simeq 17.4 \quad \rightarrow \quad f_g \simeq .025 \tag{8}$$

This suggests that the fractional free volume at the glass transition temperature is of the order of 2.5% for most known materials. The value 51.6, the second so-called universal constant, suggests, in turn that the fractional free volume divided by the free volume expansion coefficient is equal to 51.6. With a value of f_g of .025, this gives a value of the free volume expansion coefficient of about 4.8×10^{-4} per degree (Equation 9).

$$\frac{f_g}{\alpha_f} \simeq 51.6 \quad \rightarrow \quad \alpha_f \simeq 4.8 \times 10^{-4} \text{ }^\circ K^{-1} \tag{9}$$

It should be recalled that both of these parameters are nearly universal, and are valid for a very wide range of materials, although by no means for all glass-formers. This development forms the core of the so-called free volume theory which suggests that the free volume at the glass transition temperature is of the order of $2\frac{1}{2}\%$ for most materials. This concept is found to be very useful in predicting various effects, for example, those of molecular weight, plasticizer content, etc... Several other approaches to free volume theories exist (18), but the above presentation is perhaps the most instructive.

Perhaps the most sophisticated example of a thermodynamic theory is that due to Gibbs and Di Marzio (19), which concerns itself with a configurational entropy of a polymer as a function of temperature. The theory suggests that the configurational entropy approaches 0 at temperatures above 0 K, which implies that a thermodynamic glass transition should exist; this transition is called T_2. The suggestion is made that the behaviour of the experimental glass transition, as observed under normal kinetic conditions, is very similar to that of T_2. Thus, in spite of the fact that it is impossible to reach T_2, one can draw a range of conclusions in regard to the glass transition temperature from the theory of this thermodynamic T_2 transition. Within this framework, a wide range of predictions can and have been made in regard to the behaviour of the glass transition as a function of crosslink density, plasticizer content, molecular weight, etc... The theory is most illuminating but, due to its mathematical complexity, will not be discussed further.

Several kinetic theories of the glass transition have been proposed. Rather than describe any of them in detail in this short presentation, only a conceptual approach will be given which is based on the rate of volume contraction of the type shown in Figure 3. A very simple assumption concerning the rate of volume shrinkage that can be made is that it is a first order process. As expressed in Equation 10,

$$\frac{dV}{dt} = -\frac{1}{\tau_V}(V_t - V_\infty) \tag{10}$$

it is assumed that the rate of volume contraction is proportional to the volume at time t, minus the volume at infinity. The more excess free volume there is in a polymer, the faster it will shrink, the shrinkage being indicated by the minus sign, and the rate constant by $1/\tau_V$. This is not a very accurate description. First order kinetics are not exactly obeyed, and τ_V, which is the reciprocal of the first order rate constant, is dependent on the time itself. A better approximation is given in Equation 11,

$$\frac{dV}{dt} = -\frac{1}{b + at}(V_t - V_\infty) \tag{11}$$

which suggests that the rate of volume shrinkage is proportional to $1/(b + at)$, a time dependent rate constant, times $V_t - V_\infty$. a and b in this equation are constants which can be determined experimentally from measurements of the rate of volume transition in the vicinity of the glass transition. Once these constants a and b have been determined, the time dependent τ_v, the volume relaxation time, that is, the time needed for $1/2.72$ of the excess free volume to be excluded, can be determined.

The results of the approach described here applied to polystyrene are shown in Table II, tabulated as the relaxation time at various temperatures.

Table II. Volume Relaxation Times at Various Temperatures

T°C	100	95	91	90	89	88	85	79	77
τ	10^{-2} sec	1 sec	40 sec	2 min	5 min	18 min	5 hrs	60 hrs	1 yr

At approximately 100°C, relaxation time is of the order of 10^{-2} sec; at 95°, it is approximately 1 second; increasing progressively until it reaches 60 hours at 79°, and approximately one year at 77°. The glass transition for this particular sample has been determined, presumably at a cooling rate of 1°/min., as approximately 90°. It is clear, therefore, that the glass transition temperature is physically observed at a point at which the volume relaxation time is of the order of 1 to 5 minutes.

This can best be understood in the following way; if we assume that the sample is cooled at a constant rate of approximately 1° per minute, as is done in most experiments, we can be sure that volume equilibrium will be reached if the volume relaxation time is very much shorter than one minute. At very high temperatures, well in excess of 100°C, the volume relaxation time is short, of the order of nanoseconds or microseconds at high enough temperatures. This means that, for example, between 100° and 99°, we give the sample one minute to cool, while the volume relaxation time is of the order of 10 milliseconds. In the first 10 milliseconds of this cooling process, $1/2.7$ of the excess free volume will have been squeezed out, and in the next 10 milliseconds, again $1/2.7$ of the remaining excess free volume will have followed. After three or four of these relaxation time periods, that is, after thirty or forty milliseconds, almost all the excess free volume that is present in the sample, by virtue of its temperature, will have been squeezed out; however, we are giving the sample one minute to go from 100 down to 99, while the sample requires only about thirty or forty milliseconds to squeeze out all the free volume consistent with this temperature decrease. Therefore, we can be absolutely sure that when we reach 99°C, we will be at the equilibrium free volume.

The same is true in going from 95 to 94; we are giving the sample one minute for that one degree temperature change while the volume relaxation time is of the order of one second, so that in a matter of four or five seconds thermodynamic equilibrium will be reached. We can again be sure that we are at equilibrium. However, in going from 90 to 89, we again give the sample one minute; the volume relaxation time at that point, however, is two minutes. Therefore, the sample just simply does not have enough time to reach volume equilibrium. A certain amount of free volume is squeezed out, but not enough for the sample to reach equilibrium. If we kept the sample at 89° for a long period of time it would continue squeezing out free volume, but we keep lowering the temperature on the sample at the rate of 1° per minute, so that before it has even gotten rid of all the excess free volume consistent with the temperature range from 90 to 89, it is already at 88°, or 87°, or 86°, or even 85°, and it never will have the chance of getting rid of that free volume because by the time it has reached, say 77°C, the volume relaxation time is of the order of one year, but again we give it only one degree in one minute to reach the temperature of 76. Thus, a certain amount of excess free volume has been frozen in, approximately 2.5% as suggested by the W.L.F. equation, and if the sample is stored at low enough temperatures, it can never get rid of this excess free volume. However, if it is stored close to the glass transition temperature, say T_g-5 or T_g-10, that is, temperatures at which the volume relaxation times are of the order of minutes or hours, then at least some of the excess free volume can be gotten rid of, and this is observed as a volume shrinkage. This type of consideration explains the phenomena observed in Figure 3. What has been presented here is not really a kinetic theory of the glass transition but a kinetic approach to the glass transition. Several formal kinetic theories exist and the readers are referred to the reviews or the original literature (20).

In this section we consider the question of why a particular polymer has a particular glass transition temperature, or viewed another way, why the glass transition temperature of polystyrene differs, for example, from that of polypropylene. Several

MOLECULAR PARAMETERS AFFECTING THE GLASS TRANSITION

different factors which affect the glass transition temperature will be considered starting with the chain stiffness.

The mobility of polymer chains is primarily affected by the barrier to rotation around backbone carbon-carbon bonds. This, in turn, is determined primarily by the size of the substituent groups on the carbon atoms. For example, as shown in Table III, if the substituent group is a methyl group, the barrier to rotation around the carbon-carbon bond is relatively low. Thus, the glass transition temperature of the material which contains methyl groups on every second carbon atom, that is polypropylene, is -10°C. If, instead of a methyl group on every second carbon atom, a phenyl ring is placed in the same position, that is, if one looks at polystyrene, one finds that the glass transition temperature has risen to +100°C, a jump of about a 110° in changing from a methyl to a phenyl group on alternate carbon atoms of the main chain. If the substituent ring is further enlarged to (ortho)-methyl benzene, the glass transition temperature of this new material (ortho methyl styrene) has gone up to 115°C. A naphthyl group attached in the alpha position as a substituent on every second carbon atom raises the glass transition temperature to 135°C (for alpha vinyl naphthalene). Finally, for vinyl biphenyl, the glass transition temperature has gone up to 145°C. This series shows clearly that the larger the substituent, (provided that the substituent is rigid), the higher the glass transition. If we place two substituents on the carbon atom, comparing for example, styrene and alpha methyl styrene, we find that the glass transition temperature goes up from the 100°C for polystyrene to 175°C for poly (alpha methyl styrene). If we attach a naphthalene ring at two positions of the backbone rather than at one, then the glass transition temperature of 135°C for the alpha substituted naphthalene has gone up to 264°C for polyacenaphthylene. Clearly, the larger the substituent, or the more hindered the rotation because of multiple anchoring, the higher the glass transition.

While this phenomenon is very clear cut, one cannot extrapolate the behaviour to flexible side chains, or to flexible pendant groups in general. Let us explore what happens if we attach flexible pendant groups, for example, alkyl side chains, to a polymer to yield series such as the acrylates, methacrylates, alpha olefins or para alkyl styrenes. Judging from the preceding discussion, we would initially expect the glass transition temperature to go up as the size of the side chain increases. However, as we see in Figure 9, the opposite takes place: the longer the substituent, i.e., the longer the side chain, the lower the glass transition temperature.

Table III. Effect of Rigid Substituents
on T_g in $(CH_2-CHX)_n$ Polymers

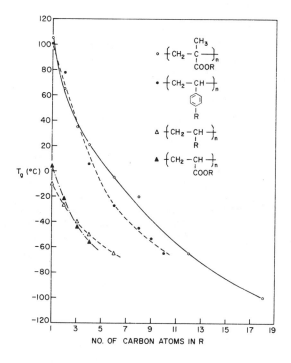

X	$-CH_3$				
T_g °C	-10	100	115	135	145
		175		264	

Figure 9. Effect of side chain length on the glass transition of
(○) methacrylates, (●) p-alkylstyrenes, (△) α-olefins, and (▲)
acrylates. (Adapted from Ref. 5.)

The reason for this phenomenon becomes clear on short deliberation. As was pointed out before, an increase in the rotational barrier increases the glass transition. However, for flexible alkyl side chains, it is only the first unit of the side chain, which is attached rigidly to the main chain, that increases the glass transition temperature. In the poly(alkyl methacrylates), only the first CH_2 group has this effect. Any additional alkyl units on the side chain would have a different effect, the reason for the difference in behaviour being that the methylene units beyond the first alkyl group can get out of the way of the rotating backbone units. Looked at another way, if we add low molecular weight materials such as hexane, octane, or decane to a polymer, we are plasticizing the polymer, because of the low T_g of the alkanes. The low molecular weight materials, as was pointed out before, have glass transition temperature of the order of -150 to -200°C. The addition of these materials as side chains to polystyrene or other polymers has the effect of depressing the glass transition temperature. Thus, the longer the side chain, the more plasticizers we are adding to a polymer, and the lower the glass transition temperature should be. This is, indeed, what is observed. As is shown quantitatively for four polymer series in Figure 9, the longer the alkyl side chain the lower the glass transition temperature.

At this point then, we have a much better appreciation for the structure of the side chain as it affects the glass transition temperature. Let us look in some detail at butyl methacrylate and inquire what happens as we change the structure of the butyl susbsituent. For t-butyl methacrylate, a bulky, and rigid group, we would expect the glass transition temperature to be highest; the secondary butyl group, which is somewhat smaller in terms of its ability to hinder rotation, should yield a lower glass transition, and a normal butyl substituent the lowest yet. This is confirmed by the data in Table IV for butyl methacrylate.

Table IV. Glass Transition Temperatures of Butyl Methacrylates (21).

CH_3 $\mid$ $-C-CH_3$ $\mid$ CH_3	CH_3 $\mid$ $-CH-CH_2-CH_3$	$-CH_2-CH_2-CH_2-CH_3$
T_g, °C 43	-22	-56

The same trends are also observed in the alpha olefins, the acrylates and the para alkyl styrenes. The behaviour is exactly what we expect, considering both the bulkiness and the internal plasticizing effect in this series of materials.

Intermolecular forces are another factor which affect the glass transition temperature profoundly. The cohesive energy density (22), which, for low molecular weight materials, is equal to the vaporisation energy divided by the molar volume, is a measure of these intermolecular forces. In polymers, since vaporisation is impossible without destruction of the chain, the cohesive energy density can be determined by swelling measurements, the cohesive energy density of the polymer being equal to that of the low molecular weight liquid which swells the polymer to the greatest extent. The polymers frequently have to be crosslinked to prevent complete dissolution in some solvents. An equation which has been proposed to correlate the cohesive energy density with the glass transition temperature is given in Equation 12.

$$CED = 0.5 \ m \ R \ T_g - 25 \ m \tag{12}$$

In this equation, m is a parameter which is analogous to the number of degrees of freedom of a molecule, or the ability of atoms or groups of atoms in the molecule to rotate. This equation is quite useful in that it illustrates the fact that intermolecular forces do indeed raise the glass transition temperature of a polymer.

Ion-containing polymers provide perhaps an ideal system for the study of intermolecular forces since the interactions between ions can be quantified most precisely. A very extensive series of studies on the polyphosphate system (23), repeat units of which are shown in Figure 10, has revealed that the glass transition temperature does, indeed, increase dramatically as the strength of the ionic interactions increases. Table V lists the glass transition temperature of a series of phosphate polymers.
HPO_3 is a nonionic polymer which has a glass transition temperature of -10°C. All the other systems are ionic. Clearly, the smaller the size or the higher the charge of the cation, the higher the glass transition temperature. The span between HPO_3 and calcium phosphate of 530°C is greater than that encompassing all the common organic synthetic polymers of commercial importance. In addition to the homopolymers described here, many mixed systems can be envisaged, that is materials in which the backbone is identical but in which some of the counterions are lithium while others are sodium, or some are calcium while others are lithium, etc...

Table V. Effect of Ionic Forces on the Glass Transition (23).

Material	T_g, °C	Pauling radius r, A	Charge q	q/a[a]
HPO$_3$	-10	-	-	0.00
LiPO$_3$	+335	0.60	1	0.50
NaPO$_3$	+280	0.95	1	0.42
Ca(PO$_3$)$_2$	+520	0.99	2	0.84
Sr(PO$_3$)$_2$	+485	1.13	2	0.79
Ba(PO$_3$)$_2$	+470	1.35	2	0.73
Zn(PO$_3$)$_2$	+520	0.74	2	0.93
Cd(PO$_3$)$_2$	+450	0.97	2	0.84

[a] a = cation radius + oxygen anion radius.

It is worth inquiring how the glass transition temperature is specifically related to the various parameters of the ions. One begins by taking the glass transition temperature as an isokinetic point, that is, a point at which the rate of the squeezing out of the excess free volume is constant. In terms of molecular mobility, this must mean that the rate of molecular motions which determines the squeezing out of the free volume must be constant, and from the specific materials under consideration here, we can conclude that this must be determined by the strength of the interactions between anions and cations. Thus, the rate determining step in these materials is considered to be the removal of the anion from the coordination sphere of the cation, or vice-versa. As shown in Equation 13, the glass transition should be proportional to the electrostatic work necessary to remove an anion from the coordination sphere of the cation. That, in turn, is proportional to the

$$T_g \quad \alpha \quad W_{el} \qquad (13)$$

electrostatic force integrated over the distance from the distance of closest approach, a, to infinity (obviously, the removal is not to infinity, but the mathematics are very much simpler if the upper limits is taken as infinity). The electrostatic force, in

$$T_g \quad \alpha \quad \int_a^\infty F_{el} \, dx$$

turn, is equal to the anion charge times the cation charge divided by the square of the distance between them.

$$T_g \quad \alpha \quad \int_a^\infty \frac{q_a q_c}{x^2} \, dx$$

which, on integration, yields

$$T_g \quad \alpha \quad \frac{q_a q_c}{a}$$

If we lump q_a, the anion charge, into the proportionality constant, we end up with Equation 14.

$$T_g \quad \alpha \quad \frac{q_c}{a} \tag{14}$$

Figure 11 shows the results of studies on the silicates, the phosphates, and the acrylates (24), the plot being one of the glass transition temperature against the cation charge, q_c, divided by the distance between centers of charge, a. Linear relationships, indeed, are observed for all three polymer systems.

For the organics, the situation is more complicated (25); Figure 12 shows the results for the ethylacrylate acrylic acid copolymer system both in the acid form and in the form of various salts. Several families of curves are observed, all of which are sigmoidal except for that corresponding to the acid. For the salts, one sees an initially rapid rise of the glass transition temperature with anion content, followed by a sigmoid with an accelerated rise of the glass transition, and then another relatively lower rate of increase at still higher ion concentrations. The onset of the sigmoid is related to the structural changes which occur in ionomers as a function of ion concentration, in this specific case the onset of cluster dominated behaviour. At very low ion contents, the ions are present primarily in the form of small aggregates, called multiplets, with the higher ion contents being characterized by the additional presence of clusters, or larger ionic aggregates. The onset of the sigmoid coincides with a point at which the clusters begin to dominate a wide range of physical properties. As might be expected, if the glass transition temperature is plotted not against the ion concentration itself, but against the normalized ion concentration, that is, the ion concentration multiplied by the q/a factor, which has been found so useful for the inorganic systems, we find that all of these lines coalesce, as shown in Figure 13, into a single line which is valid for the entire family of ethyl acrylate-acrylic acid copolymer salts.

In general, in low ion content regions, the rate of change of the glass transition with ion concentration lies anywhere between 2 and 10°C per mole % (Equation 15). The lower the T_g of the matrix, the higher the charge and the smaller the size of the cation, the greater the effect.

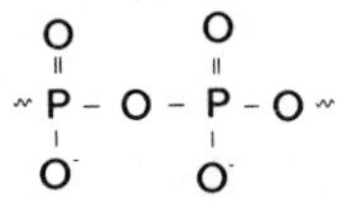

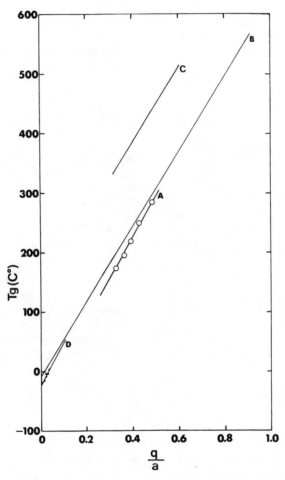

Figure 10. *The polyphosphate chain.*

Figure 11. T_g *vs.* q/a *for (A) polyacrylates, (B) polyphosphates, and (C) polysilicates. (Adapted from Ref. 24.)*

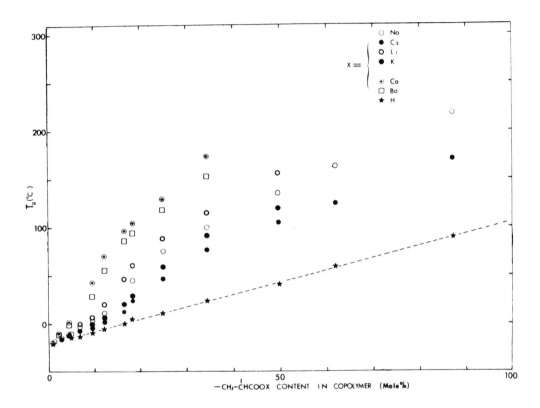

Figure 12. Glass transitions of ethyl acrylate-acrylic acid copolymers neutralized with various cations vs. ion content. (Adapted from Ref. 25.)

$$\frac{2°C}{mol \%} < \frac{\partial T_g}{\partial C} < \frac{10°C}{mol \%} \tag{15}$$

Thus, for magnesium methacrylate in polyethylene the effect will be expected to be larger than for cesium acrylate in polystyrene. A list is shown in Table VI. As can be seen, ionic interactions in polymers lead to major effects on the glass transition.

Table VI. $\frac{\partial T_g}{\partial C}$ For Various Ionomers ([26]).

Nonionic component	Ionic component	Conc. range (mole %)	$(T_g)_0$ (°C)	dT_q/dc (°C/mole %)	Ref
"Phenoxy"	Ca(SCN)$_2$	0–20	∿100	1.8	45
PPO	ZnCl$_2$	12–32	–65	4.0	47
PPO	LiClO$_4$	0–10	–70	5.5	44
Poly-sulfone	Sulfone-So$_3^-$Na$^+$	0–100	≈175	1.3	57
EA	NaA	0–12	–20	2.7	41
S	NaMA	0–10	100	3.2	28
B	LiMA	0–8	–90	5.4	15
E	NaMA	0–3	–20	5.7	35
B	MVPI	0–8	–90	8.9	15
E	Mg$_{1/2}$MA	0–2	–20	9.7	35

CONTROLLABLE PARAMETERS AFFECTING THE GLASS TRANSITION

In this section, we shall consider how the glass transition temperature of a particular polymer is affected by various parameters such as pressure, molecular weight, and diluent concentration. As mentioned before, the application of the simple free volume theory provides a useful semi-quantitative guide to the phenomena.

Within the free volume approach, if one considers what happens to a polymer above its glass transition temperature when it is subjected to hydrostatic pressure, one obtains a qualitative prediction of the effect ([27]). As one would expect, the free volume is squeezed out when pressure is applied above Tg; therefore, since the free volume is lower, the polymer is closer to its glass transition temperature.

One can also speak of a glass transition pressure at constant temperature. For example, as shown in Figure 14, one can look at the glass transition phenomenon as a function of pressure; thus, one can perform volume-pressure experiments at constant temperature, and obtain the glass transition pressure at various temperatures. Alternately, one can determine glass transition temperatures at various

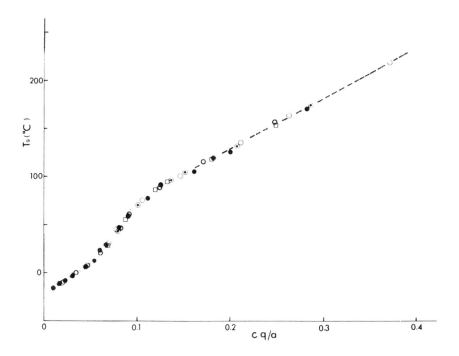

Figure 13. Glass transitions for same polymers as in Fig. 12 plotted against cq/a. (Adapted from Ref. 25.)

constant pressures. One finds that, typically, the glass transition temperature increases with pressure at the rate of approximately 20°C per thousand atmospheres, as is shown in Figure 15 for a range of materials. Thus, the pressure effect is a realistic problem when one searches for polymers for high pressure applications, for example in applications at the bottom of the ocean or other environments where pressures of the order of few hundred or few thousand atmospheres are involved. For small pressure changes of the order of one atmosphere or so, the effects are clearly negligible.

As one might expect, the presence of diluents or plasticizers decreases the glass transition temperature, an example being shown in Figure 16 for polystyrene (28). Various plasticizers have different effects on the glass transition; mostly they depress the T_g, although some additives are known which raise the glass transition temperature.

As before, the free volume approach provides an excellent semi-quantitative treatment of the plasticizer effect. Equation 16 describes the fractional free volume as a function of temperature, suggesting, as before, that the fractional free volume is equal to .025 (the value at the glass transition temperature) plus the free volume expansion coefficient αf times $(T - T_g)$.

$$f_T = 0.025 + \alpha_f(T-T_g) \qquad (16)$$

This equation is applicable both to the polymer and the diluent. If we use subscripts p and d for polymer and diluent, respectively, and for the volume fraction of polymer or diluent, V_p or V_d, then Equation 17 describes the fractional free volume at any temperature T for a polymer - diluent system.

$$f_T = 0.025 + \alpha_{fp}(T-T_{gp})V_p + \alpha_{fd}(T-T_{gd})V_d \qquad (17)$$

At the glass transition temperature, f_T becomes .025 and T becomes equal to T_g. Rearranging the equation, we obtain Equation 18 which shows the behaviour of T_g as a function of plasticizer content ($\underline{29}$).

$$T_g = \frac{\alpha_{fp} V_p T_{gp} + \alpha_{fd}(1-V_p)T_{gd}}{\alpha_{fp} V_p + \alpha_{fd}(1-V_p)} \qquad (18)$$

In this equation, T_{gd} and α_{fd}, the glass transition temperature of the diluent and the free volume expansion coefficient of the diluent, are not usually available, and sometimes not even measurable experimentally because of crystallization. If they are taken as adjustable parameters, the fit of the equation can be excellent. Many other quantitative

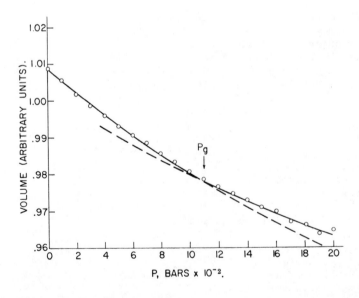

Figure 14. Volume vs. pressure for selenium at 40 °C. (Adapted from Ref. 27.)

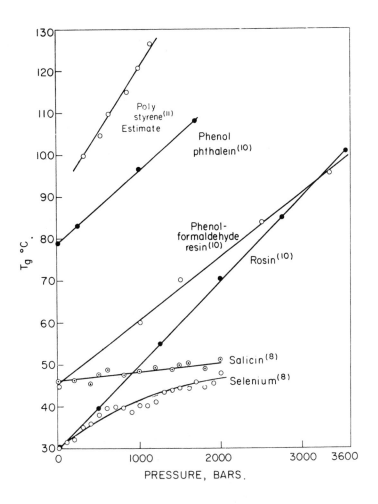

Figure 15. T_g vs. P for various materials. (Adapted from Ref. 27.)

treatments of the glass transition temperature as a function of plasticizer content are available, and the reader is referred to the original literature or the reviews (30-33).

The molecular weight is another parameter which affects the glass transition temperature of polymers. Many systems have been studied, with similar results. Again, the free volume approach gives a most useful quantitative introduction to the phenomena. It is reasonable that each chain end, at any temperature, moves a little bit more rapidly than a chain middle, since a chain end has only one "chain" attached to it, while a chain middle has two "chains" attached to it. By virtue of its greater mobility, a chain end therefore has a slightly greater excess free volume. For θ equal to the excess free volume per chain end, 2θ then becomes the excess free volume per chain, and 2θ Nav becomes the excess free volume per mole of chains. If we divide this by M, the molecular weight of the polymer, we obtain the excess free volume per gram of chains. This is summarized in Equation 19.

$$\theta = \text{excess free volume per chain end}$$
$$2\theta = \text{excess free volume per chain}$$
$$2\theta \text{ Nav} = \text{excess free volume per mol of chains} \quad (19)$$

$$\frac{2\theta \text{ Nav}}{M} = \text{excess free volume per gram of chains}$$

Finally, multiplication of this number by the density gives us the excess free volume per cubic centimeter of chains (Equation 20). This is the excess free volume (per cubic centimeter of chains) that is introduced if a chain of initially infinite molecular weight is cut down to a molecular weight of M

$$\frac{2\theta \text{ Nav}}{M}\rho = \text{excess free volume per cm}^3 \text{ of chains} \quad (20)$$

Thus, if one starts out at the glass transition temperature for the infinite polymer, one has a fractional free volume, f_g, of .025. Cutting of the chains introduces more free volume. If we now wish to get back to the glass transition temperature, this free volume must be gotten rid of by cooling the entire polymer sample to the glass transition temperature of the polymer of molecular weight M. For one cubic centimeter of material, the total free volume which must be squeezed out is simply the free volume expansion coefficient times the difference between the glass transition temperature of the infinite polymer and the glass transition temperature of a polymer of molecular weight M. As Equation 21 suggests, the total amount of free volume that was introduced per cubic centimeter of chains as a

$$\frac{2\rho\theta \; Nav}{M} = \alpha_f[T_g(\infty) - T_g(M)] \tag{21}$$

result of cutting the infinite polymer down to a molecular weight of M must be equal to the free volume expansion coefficient times the difference between the glass transition temperature of the infinite polymer and that of the polymer of molecular weight M.

Rearranging Equation 21 gives us Equation 22,

$$T_g(M) = T_g(\infty) - \frac{2\rho\theta \; Nav}{\alpha_f \; M} = T_g(\infty) - \frac{K}{M} \tag{22}$$

which shows that the glass transition temperature of a polymer of molecular weight M is equal to the glass transition temperature of polymer of infinite molecular weight minus K/M where K is simply $2\rho\theta Nav/\alpha_f$. Thus, a plot of the glass transition temperature versus the reciprocal molecular weight should give a linear relationship with a slope K and intercept of $T_g(\infty)$, as is shown in Figure 17. Since in this constant K the only unknown is θ, the excess free volume per chain end, we can calculate θ if K is known. These calculations reveal that θ is of the order of the volume of a repeat unit, that is, 20 to 50 cubic Angstroms per chain. This value can be considerably smaller if, for example, the chain end contains ionic groups which show a higher degree of attraction for each other than do chain middles in the absence of ionic interactions. A recent study of the molecular weight effect is given in reference (34).

The introduction of a comonomer into a polymer is another method of affecting the glass transition temperature. Usually, a random copolymer of two monomers with different T_g's has an intermediate glass transition temperatures. These random copolymers behave like the regular polymers in that they exhibit only one glass transition temperature. If one studies this glass transition as a function of comonomer concentration, in ideal situations one finds very simple relations between the T_g's of the homopolymers and those of the copolymers, one example being given in Equation 23 (35), where the subscripts 1 and 2 refer to pure polymers 1 and 2.

$$\frac{1}{T_g} = \frac{W_1}{T_{g1}} + \frac{W_2}{T_{g2}} \tag{23}$$

Many more complicated equations exist, examples being given in equations 24 and 25.

$$T_g = \frac{T_{g1} + k(T_{g2} - T_{g1})W_2}{1 - (1-k)W_2} \tag{24}$$

$$\frac{1}{T_g} = \frac{1}{W_1 + RW_2} \left[\frac{W_1}{T_{g1}} + \frac{RW_2}{T_{g2}}\right] \tag{25}$$

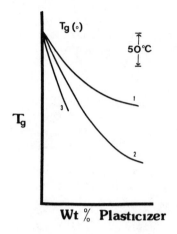

Figure 16. Glass transition temperature of polystyrene vs. plasticizer content for (1) β-naphthyl salicylate, (2) nitrobenzene, and (3) carbon disulfide. (Adapted from Ref. 28.)

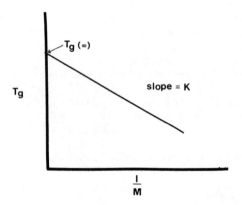

Figure 17. T_g vs. 1/M. A recent study of the molecular weight effect is given in Reference 34.

A complete listing of copolymer T_g equations is considerably beyond the scope of the present discussion. More recent studies are given in reference (38).

It is worth noting that both positive and negative deviations from ideality can be found in copolymer systems. Thus, if one takes a copolymer the homopolymers of which have identical glass transition temperatures, one can, on occasion, find negative or positive deviations, that is systems in which the glass transition temperatures of copolymers are lower or higher than those of either of the homopolymers, the crucial factor here being the rotational barriers between AB monomer pairs as compared to AA or BB monomer pairs.

The above discussion has centered on random copolymers. If the copolymers are not random, many complications can arise, the most extreme example occurring in block copolymers, that is, copolymers in which one sequence of a homopolymer A is chemically attached to one or two sequences of pure homopolymer B. If the sequences are incompatible, the materials will form phase separated polymer systems in which two glass transition temperatures are observed. This is also true of blends of two homopolymers. In this sense, the glass transition temperature can be regarded as one of the tests of compatibility for blends or blocks. If one finds one glass transition temperature, then there is a strong possibility that the systems are compatible, provided that the glass transition temperatures of the homopolymers are far enough apart. If one finds two glass transition temperatures, then presumably the systems are not compatible. Intermediate situations suggest intermediate degrees of compatibility. In this connection, the study of dynamic mechanical properties, specially the loss tangents, are quite useful.

Crosslinking is still another method of influencing the glass transition. A crosslink has the opposite effect of a decrease in molecular weight: as the crosslink density increases, the free volume of a sample decreases and glass transition temperature increases correspondingly. This is the ideal aspect; normally, the introduction of crosslinks into a polymer system occurs not just by removal of two hydrogen atoms from the chain backbone and attachment of the resulting free radicals. Crosslinking is usually accomplished by the addition of a specific crosslinking agent, which can be considered a comonomer in addition to being a crosslink. Therefore, one has two different effects to consider: one is a copolymer effect, resulting from the incorporation of a second unit, the other one a crosslinking effect. Both of these are taken into account in Equation 26 (39).

$$T_g = T_g \ (\infty) - \frac{K}{M} + K_x \ \rho \qquad (26)$$

This equation suggests that the glass transition temperature of the crosslinked polymer is equal to the glass transition temperature of the infinite polymer, minus K/M (which is used here in exactly the same sense as it was in connection with molecular weight effects) to which the $K_x \rho$ product is added; K_x being a constant and ρ the number of crosslinks per gram. Other equations have been proposed but this one illustrates the complexities involved in the study of crosslinking effects. Other examples are given in reference (40).

Crystallinity also affects the glass transition temperature, but here the results are by no means clear cut. In some cases, for example in poly-(ethylene terephthalate), as the crystallinity increases from 2 to 65%, the glass transition temperature increases from 81 to 125°C. On the other hand, in poly(4-methyl pentene), as the crystallinity increases from 0 to 76%, the glass transition temperature has been found to drop from 29 to 18°C. Many factors are responsible for this type of behaviour, including changes in tacticity with polymers which can be crystallized to different extents, molecular weight effects, etc..., and it is the effect of these secondary parameters on the degree of crystallinity which in turn influences the effect of the degree of crystallinity on the glass transition temperature.

The tacticity of the polymers is still another way of affecting the glass transition; most polymers which possess only one substituent on every second carbon atom do not show any tacticity effects; however, polymers such as poly(methyl methacrylate), which contain both a methyl group and a pendant ester on the same carbon atom, do show profound effects of tacticity. Thus, it has been observed, that syndiotactic poly(methyl methacrylate) has a glass transition temperature of 115°C, while the isotactic material has a glass transition temperature of 45°C. This phenomenon is quite general for unsymmetrically disubstituted polymers. It is not observed, as was pointed out, in monosubstituted species, for example the acrylates or the styrenes.

Still another phenomenon which affects the glass transition is the degree of elongation or the percentage of strain. Two conflicting effects are in operation here: on extension there is a slight increase in the free volume which would tend to depress the glass transition temperature; on the other hand, extension decreases the entropy of the chains, which is equivalent to increasing the glass transition temperature. Both of these effects have been observed in different cases. The effects are not large, but they are present.

A wide range of methods have become available during the last three decades for the determination of the glass transition temperature. Only a few can be mentioned here. Dilatometry is perhaps the most widely accepted classical method for the determination of the glass transition temperature, and all the early determinations were performed by that method. A dilatometer consists of a glass bulb with an attached small capillary. A sample of the polymer is placed into the large bulb, as shown in Figure 18 (left), and mercury or some other confining liquid is introduced up to a convenient level of the capillary. The dilatometer is then placed in a constant temperature bath, and the temperature changed as desired, so that a plot of volume as a function of temperature is obtained. An inflection point indicates the position of the glass transition. If this is done carefully and quantitatively, the specific volume as a function of temperature of the polymer can be obtained.

> METHODS OF DETERMINING THE GLASS TRANSITION

The procedure described above is quite laborious; a far simpler method involves the determination of one dimension of the sample only in a so-called one dimensional dilatometer, shown in Figure 18 (right). In this embodiment, the sample is placed at the bottom of a quartz tube which is attached at the other end to a linear variable differential transformer (LVDT). A quartz rod is placed on top of the sample, and to the other end of the quartz rod, in the middle of the LVDT, is attached the core of that linear variable differential transformer. As the sample expands, the core of the LVDT is moved relative to the windings of the variable differential transformer, and the length of the sample is thus translated into a voltage. Extremely high resolution is possible, and very precise determinations of the expansion coefficient can be performed by this method, using calibrated instrumentation. Both of these methods depend on the volume of the sample, and many variations on this theme are possible, including measurements of the refractive index, the transmission of ionizing radiation, for example beta rays, and others (8).

As was pointed out before, the specific heat changes as one proceeds through the glass transition in a polymer, and this aspect has become the basis of one of the most convenient methods of the determination of the glass transition, notably differential thermal analysis or differential scanning calorimetry (8). In differential thermal analysis, in principle, two containers are placed in an identical environment,

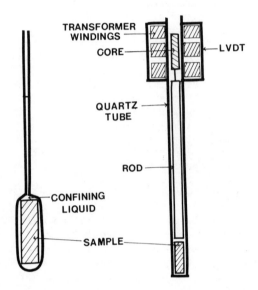

Figure 18. Volume dilatometer (left) and linear dilatometer (right). A compilation of methods is given in the introductory section in Reference 8.

for example in a metal block; one container is filled
with a polymer to be measured, the other container is
filled with some inert material, such as sand, which
does not experience any transitions in the temperature range of interest. Thermocouples are placed to
each of the containers, and the temperature of the
aluminum block, which contains both of these containers, is changed. If the temperature is changed at
a constant rate, a discontinuity in any thermal
property in one of the samples will be manifested as
a change in the temperature difference between the
two samples. Thus, since the heat capacity changes
as one proceeds through the glass transition temperature of a polymer, and since no comparable change has
occurred in the blank (or the sand) sample, one will
observe a change in the ΔT, the temperature
differences between the sand and the polymer, as one
proceeds through the glass transition temperature of
the polymer. This is an extremely convenient method,
suitable for automation, and several commercial
instruments are available which accomplish this. A
slight modification of this method is the so-called
differential scanning calorimetry method, which
measures the amount of heat that must be supplied to
one or the other of the two samples to keep them at
a constant temperature. Thus, one can measure exo-
thermic or endothermic phenomena in the sample.
Again, commercial instruments are available and the
method is rapid and completely automatic.

MOLECULAR MOTIONS BELOW THE GLASS TRANSITION

The last topic to be discussed in this chapter
concerns molecular motions below the glass transition
temperature. Only one example of a study will be
presented here, as a more thorough discussion is
beyond the scope of this presentation. A number of
review articles have been written on this topic, and
the reader is referred to these reviews or the
original literature (41-42).

As was mentioned in connection with Figure 8,
low temperature peaks are observed when the loss
tangent is measured in all polymers. These low
temperature peaks generally reflect a molecular
mechanism, for example a movement of a benzene ring,
of a cyclohexyl group, a methyl group, etc., depending on the type of groups that are present in the
polymer. The field of study is known as mechanical
spectroscopy (42), and it concerns itself with the
elucidation of which segment of the molecule moves
with which frequency as a function of temperature.
In a typical study, the loss tangent is measured at
a constant frequency as a function of temperature.
Various frequencies require a very wide range of
instruments, and the instrument which is used at 1 Hz,

for example a torsion pendulum, differs dramatically from that which is used in the neighbourhood of 1 KHz, for example a vibrating reed, which in turn differs drastically from the instrument which is used in the ultrasonic region at 1 MHz.

One example of a mechanical spectroscopy study will be described here. This example concerns the cyclohexyl group as investigated in various polymers by Heijboer and his coworkers (41). In this study, it was found that whenever a cyclohexyl group is attached to a polymer chain, it gives rise to peaks of the type shown in Figure 19. The investigation extended over the frequency range from 10^{-4} cycles per second all the way to 8×10^5 cycles per second, the results showing a remarkable constancy of the peak height as a function of temperature and frequency. Mechanistically, the peak was assigned to a chair-chair flip within the cyclohexyl group of the type shown in Figure 20. The activation energy for the motion was obtained from the graph of the log of the frequency at maximum loss plotted against the reciprocal temperature, that is from an Arrhenius plot, which gives a straight line consistent with a slope of 11.2 K cal. Subsequently, in an NMR investigation, the barrier for the chair-chair flip was, indeed, found to be approximately 11.2 K cal, in excellent agreement with the earlier mechanical study. It seems unimportant whether the cyclohexyl group was attached to a methacrylate polymer or a different type of polymer, or even not at all atta- ched to a polymer chain. Under most conditions, the barrier to the chair-chair flip was constant and uniquely assignable to this specific motion of the cyclohexyl group.

Since that study appeared, many other materials have been investigated by similar techniques. In some cases, as mentioned before, relations have been found between brittleness of glasses and the absence of these low temperature relaxations, but these studies are by no means conclusive, and much future work remains to be done. Suffice it to say that materials below the glass transition temperature are not, by any stretch of the imagination, rigid, unmoving materials; the molecules can perform a wide variety of motions, and if the material is close enough to the glass transition temperature, considerable volume shrinkage can occur as was seen above.

CONCLUSION

The glass transition phenomenon represents a relative- ly new field, the phenomenology having been explored only since the 1920's. While the phenomenology is well known, there is, as yet, no thorough

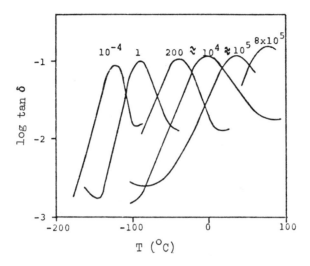

Figure 19. Log tan δ vs. temperature for poly(cyclohexyl methacrylate). (Adapted from Ref. 41.)

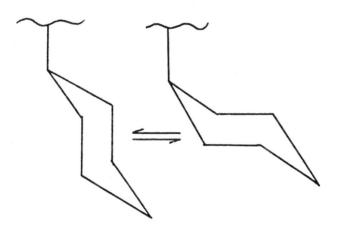

Figure 20. Chair-chair flip in pendant cyclohexane groups. An introductory review of mechanical spectroscopy is given in Reference 42.

understanding of the theoretical background of the glass transition, although attempts have been made, and many have been quite successful, at least in part. The field, however, represents one of enormous industrial importance, in that an ability to control the glass transition temperature, and to understand it, aids in the industrial utilization of polymeric materials. Finally, it was shown that materials in the glassy state are very much "alive", in that they are subject to specific molecular motions on a small scale, and that in the vicinity of the glass transition the material can shrink considerably with time.

LITERATURE CITED

Reviews, Symposia, and Tabulations

1. Tammann, Gustav "Der Glazustand"; Leopold Voss: Leipzig, 1933.
2. Kauzmann, W. Chem. Revs. 1948, 43, 219.
3. Boyer, R.F. Rubber Chem. Tech. 1963, 36, 1303.
4. Kovacs, A.J. Fortschr. Hochpolym. Forsch. 1963, 3, 394.
5. Shen, M.C.; Eisenberg, A. in "Progress in Solid State Chemistry"; Reiss, H., Ed.; Vol. 2, Chap. 9.
6. Shen, M.C.; Eisenberg, A. Rubber Chem. Tech. 1970, 43.
7. Wrasidlo, W. in "Thermal Analysis of Polymers"; Advances in Polymer Sci. 1974, 13, 3 pp. 29-53.
8. "Polymer Handbook", 2nd ed. Brandrup J.: Immergut, E.H., Eds; John Wiley and Sons: New York, 1975, pp. III-139 to III-192.
9. "The Glassy Transition and the Nature of the Glassy State"; Goldstein M.: Simha, R., Eds. Annals of the N.Y. Acad. Sci. 1976, 279.
10. "Structure and Mobility in Molecular and Atomic Glasses"; O'Reilly, J.M.: Goldstein M., Eds. Annals of the N.Y. Acad. Sci. 1981, 371.

Specific

11. Jones, G.O. "Glass"; Methven: 1956.
12. Tobolsky, A.V. "Properties and Structure of Polymers"; John Wiley: 1960.
13. Cayrol, B.; Eisenberg, A.; Harrod, J.F.; Rocaniere, P. Macromolecules 1972, 5, 676.
14. "The Glassy Transition and the Nature of the Glassy State"; Goldstein M.: Simha, R., Eds. Annals of the N.Y. Acad. Sci. 1976, 279.
15. "Structure and Mobility in Molecular and Atomic Glasses"; O'Reilly, J.M.: Goldstein M., Eds. Annals of the N.Y. Acad. Sci. 1981, 371.
16. Doolittle, A.K. J. Appl. Phys. 1951, 22, 1471.
17. Williams, M.L.; Landel, R.F.; Ferry, J.D. J. Am. Chem. Soc. 1955, 77, 3701.

18. "Structure and Mobility in Molecular and Atomic Glasses"; O'Reilly, J.M.: Goldstein M., Eds. Annals of the N.Y. Acad. Sci. 1981, 371, p. 199.
19. "Structure and Mobility in Molecular and Atomic Glasses"; O'Reilly, J.M.: Goldstein M., Eds. Annals of the N.Y. Acad. Sci. 1981, 371, p. 21
20. "Structure and Mobility in Molecular and Atomic Glasses"; O'Reilly, J.M.: Goldstein M., Eds., Annals of the N.Y. Acad. Sci. 1981, 371.
21. Shetter, J.A. Polymer Letters 1963, 1, 209.
22. Hayes, R.A. J. Appl. Poly. Sci. 1961, 5, 318.
23. Eisenberg, A.; Farb, H.; Cool, L.G. J. Poly. Sci. A-2 1966, 4, 855.
24. Eisenberg, A.; Matsuura H.; Yokoyama, T. J. Poly. Sci. A-2 1971, 9, 2131.
25. Matsuura, H.; Eisenberg, A. J. Polymer Sci. Polymer Phys. Ed. 1976, 14, 1201.
26. Eisenberg, A.; King, M. "Ion-Containing Polymers", Acad. Press: N.Y., 1977.
27. Eisenberg, A. J. Phys. Chem. 1967, 67, 1333.
28. Jenckel, E.; Heusch, R. Koll. Z. 1953, 130, 89.
29. Kelley, F.N.; Bueche, F. J. Polymer Sci. 1961, 50, 549.
30. Couchman, P.R.; Karasz, F.E. Macromolecules 1978, 11, 117.
31. Chow, T.S. Macromolecules 1980, 13, 362.
32. Czekaj, T.; Kapko, J. Europ. Polymer J. 1981, 17, 1227.
33. ten Brinke, G.; Karasz, F.E.; Ellis, T.S. Macromolecules 1983, 16, 244.
34. Couchman, P.R. Polymer Engineering and Science 1981, 21, 377.
35. Fox, T.G. Bull. Am. Phys. Soc. 1956, 1, 123.
36. Gordon, M.; Taylor, J.S. J. Appl. Chem. 1952, 2, 493.
37. Mandelkern, L.; Martin, G.M.; Quinn, F.A. J. Res. NBS 1957, 58, 137.
38. Couchman, P.R. Nature 1982, 268, 729. also Macromolecules 1982, 15, 770.
39. Fox, T.G.; Loshaek, S. J. Polymer Sci. 1955, 15, 371.
40. Loshaek, S. J. Poly. Sci. 1955, 15, 391.
41. Heijboer, J. Ph.D. Thesis, Univ. Leiden, Netherlands, 1972.
42. Eisenberg, A.; Eu, B.C. Ann. Revs. Mat'l. Sci. 1976, 6, 335.

3

Viscoelasticity and Flow in Polymer Melts and Concentrated Solutions

William W. Graessley

This part of the course deals with viscoelasticity, particularly as it relates to the flow properties of polymer melts and concentrated solutions. Professor Eisenberg has discussed the glassy state and the glass transition in polymer liquids. The concern here, however, is with the response at long times and at temperatures that lie well above the glass transition. Professor Mark has discussed the elastic behavior of polymer networks well above the glass transition. The conditions here are similar, but the chains are not linked to form a network. All of the molecules have finite size, and since the discussion is limited to flexible chain polymers, they have random coil conformations at equilibrium.

We will cover linear viscoelasticity, a primary means of rheological characterization for polymer liquids, and steady shear flows, a relatively well understood bridge into non-linear behavior. The effects of large scale chain structure (molecular weight, molecular weight distribution, long chain branching) will be described, and current theoretical ideas about molecular effects will be described. The quantitative aspects, which can become very technical, will not be emphasized. The primary aim is to convey some physical understanding about viscoelastic behavior in polymer liquids from both the macroscopic and molecular viewpoints.

Deformation means a change in shape. A liquid deforms as it flows in a tube (Fig. 1), driven by

0851/84/0097$14.40/1
© 1984 American Chemical Society

pressure or gravity. The particles of liquid move in straight lines parallel to the tube axis. Those on the center line have the maximum velocity. The particles at the wall do not move at all. In long tubes each particle moves at constant velocity if the driving force is constant, slightly slower than adjacent particles which are nearer the center line and slightly faster than those nearer the wall. Tube flow exemplifies <u>simple shear</u> deformation. The layers of liquid along the flow direction slide over one another without stretching. Other examples of simple shear are the flows induced by relative rotation of a pair of coaxial cylinders and a coaxial circular cone and plate.

   <u>Simple extension</u> belongs to a different class of deformations. Layers of particles along the flow direction are stretched without sliding relative to adjacent layers. Extensional flows are important in many polymer processing operations, for example in fiber spinning or film forming. They are difficult to generate and sustain in a controlled way, however. Most methods used to characterize polymer flow properties involve simple shear flows. The presumed effects of these two classes of flow on the chain conformations are sketched in Fig. 2.

   Deformation causes a relative motion which induces resisting forces. Constant particle velocities in flow through a tube (Fig. 1) are the result of a balance of two forces. The force from the pressure difference is opposed by a force from the <u>shear stress</u>, acting between the adjacent layers of deforming liquid along the flow direction. The applied torque in the concentric cylinder and cone and plate geometries is opposed similarly. <u>Stress,</u> in general, describes the force per unit area transmitted by contact between adjacent layers of particles. The relationship between stress and deformation is a property of the material. <u>Rheology</u> is the study of stress-deformation relationships.

   Liquids have no preferred shape. In a <u>perfectly viscous liquid</u> the stress, apart from the pressure which always acts equally in all directions, depends only on the <u>rate of deformation</u>. The history of deformation is irrelevant in a perfectly viscous liquid. The stress at each moment depends only on how rapidly it is being deformed at that moment. A perfectly viscous liquid has no memory. The mechanical work expended in producing the formation is dissipated instantaneously.

   Compare this with the behavior of solids. Solid bodies have a preferred shape, the shape the body assumes spontaneously when no force is applied, and in a <u>perfectly elastic solid</u> the stress depends only on the amount of deformation from that preferred shape. The mechanical work to produce the deformation is

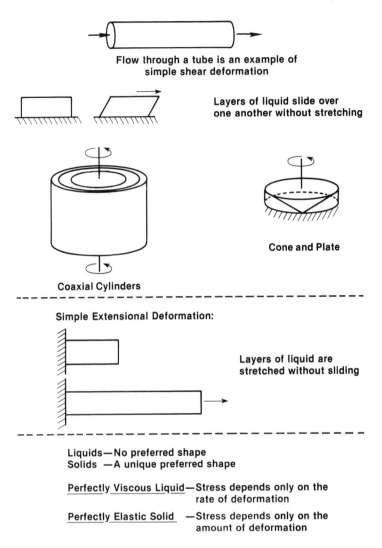

**Flow through a tube is an example of
simple shear deformation**

**Layers of liquid slide over
one another without stretching**

Cone and Plate

Coaxial Cylinders

Simple Extensional Deformation:

**Layers of liquid are
stretched without sliding**

Liquids—No preferred shape
Solids —A unique preferred shape

<u>Perfectly Viscous Liquid</u>—Stress depends only on the
rate of deformation

<u>Perfectly Elastic Solid</u> —Stress depends only on the
amount of deformation

Figure 1. Types of deformation and response of solids and
liquids.

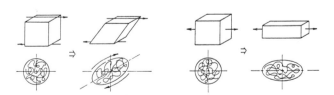

Figure 2. Shear and elongation.

stored as elastic energy. A substance is viscoelastic if it exhibits both energy dissipation and energy storage in its mechanical properties. In a viscoelastic liquid the stress depends on the history of the deformation. Some finite time must pass for the liquid to "forget" a shape that it had in the past.

All real substances are viscoelastic. How they respond in particular situations depends on the rate of spontaneous structural reorganization at the molecular level. In ordinary liquids that are well above the glass transition, the local arrangements of molecules relaxes quickly through the action of Brownian motion (Fig. 3). Structural "memory" is very short, changes in distance between molecules induced by the deformation are quickly relaxed, and the response is essentially viscous unless the testing rate is extremely rapid. In ordinary solids the relaxation of structure is extremely slow. Structural memory is very long, and the response is essentially elastic. One property that sets polymer systems apart is the enormously wide span of times over which relaxation takes place.

The spreading of relaxation over many orders of magnitude in time is a natural consequence of macromolecular structure. Rearrangement of flexible chains may be very rapid on the scale of a chain unit. That rate is affected by the nature of the chain unit itself and to some extent by the need for local cooperation along the chain, but it is probably not different in any significant way from the rearrangement rate in ordinary liquids at the same temperature relative to the glass transition. However, the full equilibration of chain conformation requires much longer times. Relaxation propagates over larger and larger chain distances with time, and the longest times are strongly dependent on the large scale chain architecture. These relatively sluggish processes, the terminal region relaxations, govern the flow properties. This is why molecular weight and long chain branching are so important in the rheological properties of polymers.

We say that polymer networks are true solids because they have a preferred shape and can support a deformed shape at equilibrium. Polymer melts and solutions are true liquids in the sense of having no preferred shape at equilibrium. However, each may require a significant time to reach the final equilibrium when deformed to a new shape. Chain configuration can be displaced significantly from equilibrium by deformation in both polymer networks and polymer liquids. Elastic effects are readily demonstrated in polymer liquids, just as in networks. This ease of evoking finite strain effects is the second property which distinguishes polymeric and monomeric liquids.

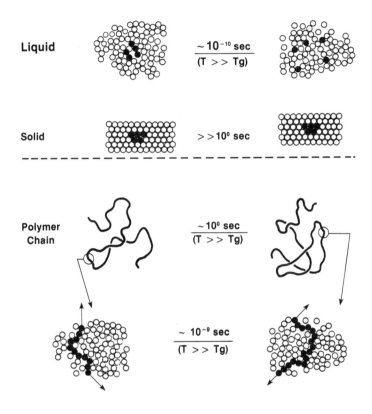

Liquid $\quad\dfrac{\sim 10^{-10}\ \text{sec}}{(T >> Tg)}$

Solid $\quad >> 10^{0}\ \text{sec}$

Polymer Chain $\quad\dfrac{\sim 10^{0}\ \text{sec}}{(T >> Tg)}$

$\dfrac{\sim 10^{-9}\ \text{sec}}{(T >> Tg)}$

- Local rearrangement of conformation is rapid

- Rearrangement over larger chain distances is slower

- Complete relaxation requires times which depend very strongly on the chain architecture

Figure 3. Time-dependent response; molecular viewpoint.

LINEAR VISCOELASTI-
CITY

If the deformation is small, or applied sufficiently slowly, the molecular arrangements are never far from equilibrium. The mechanical response is then just a reflection of dynamic processes at the molecular level which go on constantly, even for the system at equilibrium. This is the domain of <u>linear visco-elasticity</u>. The magnitudes of stress and strain are related linearly, and the behavior for any liquid is completely described by a single function of time. The properties of this function can be measured by a variety of procedures. The particular case of simple shear deformation is considered in Fig. 4.

Simple shear deformation between parallel plates is sketched at the top of Fig. 4. Shear stress σ is the shear force per unit area of the surface it acts upon (the shear plane); shear strain γ is the displacement per unit distance normal to the shear plane; shear rate $\dot{\gamma}$ is the rate of change of shear strain with time. These quantities are uniform throughout the material in the example - the deformation is homogeneous.

The stress-strain behavior for Hookean solids and Newtonian liquids, the classical models of purely elastic response for small deformation and of purely viscous response, are given at the bottom of Fig. 4. A single constant defines the mechanical response in each case, the shear modulus G for the solid and the shear viscosity η for the liquid. All that matters for the stress is the current value of strain for the solid and the current value of strain rate for the liquid. The history of loading does not enter in either case.

Such is not the case for a viscoelastic substance (Fig. 5). The response moves from solid-like at short times to liquid-like at long times, and the history of loading is crucial. In the <u>stress relax-ation</u> experiment, a constant shear strain is imposed instantaneously (in principle), and the stress at subsequent times is monitored. The stress (for t >0) is constant for a Hookean solid because the strain is constant. The stress (for t >0) is zero for a Newtonian liquid because the strain rate is zero. The stress in a viscoelastic substance decreases with time, starting at some high value and going finally to zero for a viscoelastic liquid. For small enough strains the ratio of stress to strain is a function of time alone. That function is a property of the material called the stress relaxation modulus G(t).

Shear Deformation

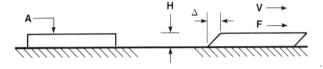

$$\frac{F}{A} \quad = \quad \frac{\text{shear force}}{\text{area}} \quad = \text{ shear stress } \sigma \text{ (dyne/cm}^2)$$

$$\frac{\Delta}{H} \quad = \quad \frac{\text{displacement}}{\text{spacing}} \quad = \text{ shear strain } \gamma \text{ (dimensionless)}$$

$$\frac{V}{H} \quad = \quad \frac{(d\Delta/dt)}{H} \quad = \text{ shear rate } \dot{\gamma} \text{ (sec}^{-1})$$

Perfectly Elastic Solid (small strains):

$$\sigma(t) = G\,\gamma(t) \qquad \text{Hooke's Law}$$

$$G = \text{ shear modulus } \text{(dyne/cm}^2)$$

Perfectly Viscous Liquid:

$$\sigma(t) = \eta\,\dot{\gamma}(t) \qquad \text{Newton's Law}$$

$$\eta = \text{ shear viscosity } \text{(dyne—sec/cm}^2; \text{ poise)}$$

Figure 4. Linear stress-strain behavior for solids and liquids.

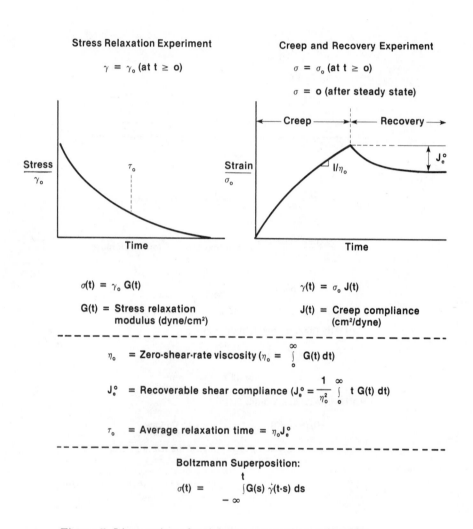

Stress Relaxation Experiment

$\gamma = \gamma_o$ (at $t \geq o$)

$$\sigma(t) = \gamma_o\, G(t)$$

$G(t)$ = Stress relaxation modulus (dyne/cm²)

Creep and Recovery Experiment

$\sigma = \sigma_o$ (at $t \geq o$)

$\sigma = o$ (after steady state)

$$\gamma(t) = \sigma_o\, J(t)$$

$J(t)$ = Creep compliance (cm²/dyne)

- -

η_o = Zero-shear-rate viscosity ($\eta_o = \displaystyle\int_o^\infty G(t)\,dt$)

J_e^o = Recoverable shear compliance ($J_e^o = \dfrac{1}{\eta_o^2} \displaystyle\int_o^\infty t\,G(t)\,dt$)

τ_o = Average relaxation time = $\eta_o J_e^o$

- -

Boltzmann Superposition:

$$\sigma(t) = \int_{-\infty}^{t} G(s)\, \dot\gamma(t\text{-}s)\, ds$$

Figure 5. Linear viscoelastic response (polymer liquids).

In $\underline{creep}$ experiments a constant stress is
imposed and the strain at subsequent times is monitored; in $\underline{creep\ recovery}$ the sample is unloaded (the
stress is reduced to zero), and the strain at subsequent times is monitored. The strain is constant for
the Hookean solid and the strain rate is constant for
the Newtonian liquid in the creep experiment (the
strain increases at a constant rate) because the
stress is constant. The strain (for the solid) and
strain rate (for the liquid) respectively go immediately to zero with unloading (the creep recovery
phase) because the stress is now zero. The strain
rate of a viscoelastic liquid decreases with time in
the creep experiment, finally reaching a steady state
where the strain rate is constant. With unloading
(the creep recovery phase) the liquid recoils, eventually reaching equilibrium at some smaller total
strain relative to the strain at unloading. If the
stress is small enough the response is linear: the
ratio of time-dependent strain to stress in the
creep phase is the creep compliance J(t), and the
total recoverable strain in the subsequent recovery
phase is proportional to the stress.

The linear viscoelastic properties G(t) and J(t)
are not independent. They reflect the same dynamic
processes at the molecular level in the equilibrium
liquid and are related through the $\underline{Boltzmann\ Super}$-
$\underline{position\ Principle}$ (Fig. 5, bottom). This is not
the place to discuss that important relationship in
detail; we will simply quote results. Two properties
which play an important role in flow behavior are the
steady state viscosity at zero shear rate η_o and the
steady state recoverable shear compliance J_e^o. They
are obtained quite directly from results of the creep
experiment, η_o from σ and $\dot{\gamma}$ in the steady state
region and J_e^o from the total recoil strain at steady
state. Boltzmann superposition gives expressions for
η_o and J_e^o in terms of G(t). It also provides an
expression for τ_o, an average relaxation time for
G(t). It is simply the product $\eta_o J_e^o$, a measure of
time required for final equilibration following a
step strain.

The interplay of the configurational distortion
produced by deformation and Brownian motion, which
relaxes that distortion, is sketched in Fig. 6. We
will return to this molecular picture later.

The behavior of G(t) for polymer melts, illustrating the glassy, transition, plateau and terminal
regions is sketched in Fig. 7. Deformation carries
the chains into distorted conformations. At very
short times the response is glassy. The modulus is
large (typically 10^{11} dynes/cm^2). It falls rapidly
as the chains relax locally, and then over progressively longer chain distances. For short chains the

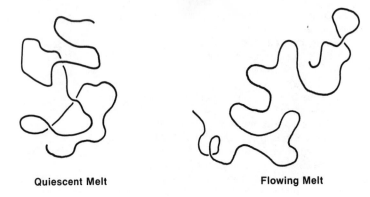

Quiescent Melt   Flowing Melt

(1) Coil distortion—Governed by competition between hydrodynamic forces and Brownian (diffusional) forces.

(2) Elastic-like properties of melts—related to the reduction in available configurations with coil distortion.

(3) Coil distortion increases with chain length.

Figure 6. Flow, molecular configuration, and viscoelasticity.

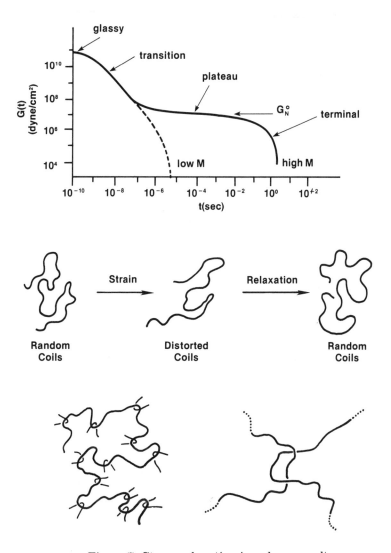

Figure 7. Stress relaxation in polymer melts.

relaxation proceeds smoothly to zero. For long chains the relaxation rate slows, and the modulus remains relatively flat for a time before finally resuming its rapid relaxation to zero. This intermediate region, or plateau, separates the short time response, called the transition region, where the chain architecture has little effect, from the long time response, called the terminal region, where such architectural features as molecular weight, molecular weight distribution, and long-chain branching have a profound effect. The response at intermediate times resembles that of a network. The width of the plateau increases rapidly with chain length, but the plateau modulus G_N^O is independent of chain length. This behavior is attributed to chain entanglement, or, more precisely, to chain uncrossability. At high polymer concentrations the domains of individual chains overlap extensively (bottom of Fig. 7). Long chains equilibrate in the deformed state up to a certain chain distance (the "entanglement spacing"), but further equilibration is delayed until the chains extricate themselves from the constraining mesh of surrounding chains. They must somehow diffuse around the contours of their constraining neighbors in order to obliterate all memory of the original shape.

Although stress relaxation and creep experiments are used extensively, the most common method to determine linear viscoelastic properties is oscillatory strain response. Figure 8 illustrates that technique. The liquid is strained sinusoidally, and the in-phase and out-of-phase components of the stress at steady state are measured as functions of the frequency ω. The strain amplitude is kept small enough to evoke only a linear response. In a perfectly viscous liquid the stress is exactly in-phase with the strain rate, which is to say, exactly 90° out-of-phase with the strain. For a viscoelastic substance the phase angle and stress amplitude vary with frequency. The in-phase component of stress defines the dynamic storage modulus $G'(\omega)$; the 90° out-of-phase component defines the dynamic loss modulus $G''(\omega)$. Results are sometimes expressed in terms of auxiliary functions such as the dynamic viscosity, $G''(\omega)/\omega$, the absolute magnitude of the complex viscosity, $[G'^2(\omega) + G''^2(\omega)]^{1/2}/\omega$, and the loss tangent $G''(\omega)/G'(\omega)$.

The sketch in Fig. 9 shows $G'(\omega)$ and $G''(\omega)$ for a high molecular weight polymer melt (see $G(t)$ for the same liquid in Fig. 7). Compared with $G(t)$, the order of appearance of the various regions is reversed. Low frequencies correspond to long times, and vice versa. At low frequencies $G'(\omega)$ is much smaller than $G''(\omega)$: the viscous response dominates. At intermediate frequencies $G'(\omega)$ is larger than

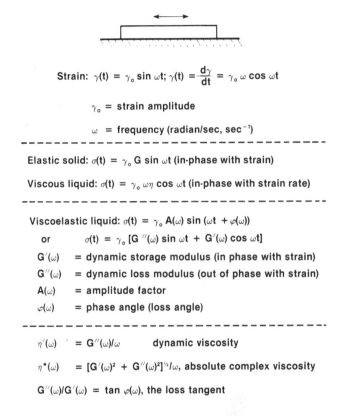

Strain: $\gamma(t) = \gamma_0 \sin \omega t$; $\gamma(t) = \dfrac{d\gamma}{dt} = \gamma_0 \, \omega \cos \omega t$

γ_0 = strain amplitude

ω = frequency (radian/sec, sec^{-1})

- -

Elastic solid: $\sigma(t) = \gamma_0 \, G \sin \omega t$ (in-phase with strain)

Viscous liquid: $\sigma(t) = \gamma_0 \, \omega\eta \cos \omega t$ (in-phase with strain rate)

- -

Viscoelastic liquid: $\sigma(t) = \gamma_0 \, A(\omega) \sin (\omega t + \varphi(\omega))$

or $\qquad \sigma(t) = \gamma_0 \, [G''(\omega) \sin \omega t + G'(\omega) \cos \omega t]$

$G'(\omega)$ = dynamic storage modulus (in phase with strain)

$G''(\omega)$ = dynamic loss modulus (out of phase with strain)

$A(\omega)$ = amplitude factor

$\varphi(\omega)$ = phase angle (loss angle)

- -

$\eta'(\omega)$ $\ \ = G''(\omega)/\omega$ $\qquad$ dynamic viscosity

$\eta^*(\omega)$ $\ = [G'(\omega)^2 + G''(\omega)^2]^{1/2}/\omega$, absolute complex viscosity

$G''(\omega)/G'(\omega) = \tan \varphi(\omega)$, the loss tangent

Figure 8. Oscillatory strain response.

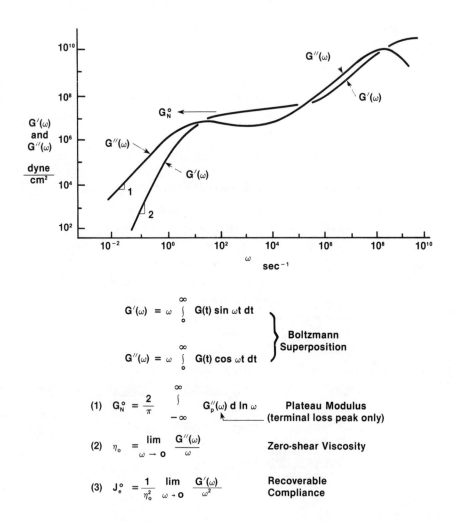

$$G'(\omega) = \omega \int_0^\infty G(t) \sin \omega t \, dt \left. \right\}$$

$$G''(\omega) = \omega \int_0^\infty G(t) \cos \omega t \, dt$$

Boltzmann Superposition

(1) $G_N^o = \dfrac{2}{\pi} \displaystyle\int_{-\infty}^{\infty} G_p''(\omega) \, d \ln \omega$ Plateau Modulus (terminal loss peak only)

(2) $\eta_o = \displaystyle\lim_{\omega \to 0} \dfrac{G''(\omega)}{\omega}$ Zero-shear Viscosity

(3) $J_e^o = \dfrac{1}{\eta_o^2} \displaystyle\lim_{\omega \to 0} \dfrac{G'(\omega)}{\omega^2}$ Recoverable Compliance

Figure 9. Dynamic moduli for polymer melts.

G"(ω) and is relatively constant. The magnitudes
reverse again in the transition region. G'(ω)
finally levels off and G"(ω) falls again in the
glassy region. Note that the loss modulus has two
peaks, corresponding to the terminal region (low
frequencies) and the transition region (high frequencies. Boltzmann superposition requires that G"(ω) become proportional to frequency and G'(ω) to the square
of frequency at low enough frequencies for any liquid
(slopes of 1 and 2 on log-log plots). The plateau
modulus, the viscosity and the recoverable compliance
can be obtained by the expressions at the bottom. Note
that the relationship between loss modulus and plateau
modulus involves integration over the terminal loss
peak alone. That equation is most useful when the two
dispersions are well resolved, meaning long chains and
narrow chain length distribution. The plateau modulus
acts, in effect, as the initial modulus for the terminal region, the part that dominates the flow behavior.

Notice that the stress relaxation and dynamic
moduli encompass many orders of magnitude in time
and frequency. No single experiment can cover the
entire range; five orders of magnitude is a typical
limit, although this can be pushed slightly further
in some cases. The sketches in fact represent master
curves - composites of data measured at different
temperatures. Most polymer liquids obey time-
temperature superposition, shown on Fig. 10. Temperature shifts the viscoelastic functions along the
modulus and time (or frequency) scales without
changing their shapes. In fact, the modulus shift
is extremely weak. The major effect of temperature
change is simply to rescale the time. Raising the
temperature shifts the response curves to smaller
times (higher frequencies). At the molecular level
the rate of Brownian motion is increased but the
molecular organization, the "structure of the liquid"
is practically unchanged. Measurements at different
temperatures can thus be reassembled to form a master
curve, covering many more decades than is possible
by measurements at any single temperature.

Typical behavior is shown in Figure 10 where
storage modulus vs. frequency is plotted at several
temperatures for a sample of polystyrene. A convenient reference temperature is selected, in this case
160°C, and "best fit" scale factors for data at other
temperatures are determined empirically. The time
scale can shift very rapidly, as seen by the plot of
a_T vs T. Note also that a_T is independent of chain
length except for very short chains. It is a property of the local composition.

The combination of time-temperature superposition and Boltzmann superposition leads to expressions

$$G(t) = G_r(t/a_T)$$
$$G'(\omega) = G_r'(\omega\, a_T)$$
$$G''(\omega) = G_r''(\omega\, a_T)$$

**Time-temperature superposition
(negligible vertical (modulus) shift)**

**"r" means the behavior at the
chosen reference temperature**

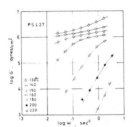

*(Reproduced from Ref. 1. Copyright
1970, American Chemical Society.)*

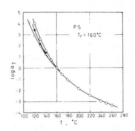

*(Reproduced from Ref. 1. Copyright
1970, American Chemical Society.)*

$$\log a_T = \frac{-7.14(T - 160)}{112.1 + (T - 160)} \qquad (2)$$

When the vertical shift is negligible:

$$G_N^\circ(T) = G_N^\circ(T_r)$$
$$\eta_o(T) = a_T\eta_o(T_r)$$
$$J_e^\circ(T) = J_e^\circ(T_r)$$

**Both G_N° and J_e° are
practically independent
of temperature**

Figure 10. Effect of temperature on viscoelastic response.

for the temperature dependence of viscosity, plateau modulus and recoverable compliance (bottom of Fig. 10). If the modulus shift is negligible, as is normally the case, the plateau modulus and recoverable compliance are practically independent of temperature and the temperature coefficient of viscosity is given directly by a_T. This is extremely useful for extrapolation purposes. Properties of polymer melts can be measured at experimentally convenient temperatures and estimated elsewhere using only a_T, the temperature shift factor for the polymer species.

Dynamic moduli master curves for narrow distribution linear polystyrenes of different molecular weights are shown in Fig. 11. Note the increase in width of the plateau with molecular weight and the similarity in shape of the terminal dispersion for different molecular weights. Note also that the plateau modulus is increasingly more well defined with increasing chain length, and that its value, $G_N^O = 2.0 \times 10^6$ dyne/cm^2 for polystyrene, is independent of chain length.

Master curves for two polystyrene samples of different molecular weight distributions are also shown in Fig. 11. We see that the samples have similar values of viscosity because their loss moduli merge at low frequencies (note Eq. (2) in Fig. 9). However, their recoverable compliances are quite different. At the same viscosity J^O depends only on $G'(\omega)$ at low frequencies (Eq. 3 in Fig. 9), and these are much larger for the sample with broad distribution. Indeed, the data for that sample have still not reached the limiting behavior ($G' \alpha \omega^2$) at the lowest available frequencies. This illustrates the general point that J_e^O is extremely sensitive to distribution, and particularly to the presence of a high molecular weight tail. Molecular weight distribution strongly affects the shape of the terminal region. The response is "smeared out", so to speak, because chains of different size relax to equilibrium at different times. In the example here, the terminal region is so broad that the peak in G'', quite well defined in the narrow distribution sample, becomes merely a broad shoulder on the transition region. The curves for the two samples merge at high frequencies. Response there depends on local chain motions; the effects of chain length and distribution are gone.

We turn now to behavior in steady simple shear flows – the shear rate dependence of viscosity, normal stress effects, and an elastic recoil phenomenon related to melt processing, die swell. These are

NON-LINEAR VISCOELASTICITY

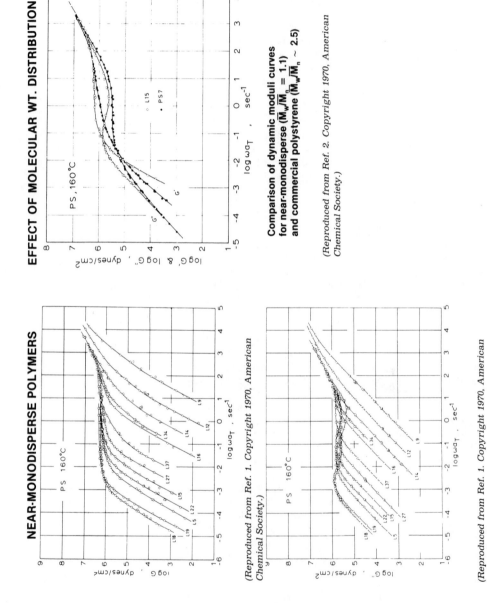

Figure 11. Effect of molecular weight distribution on near-monodisperse polymers.

non-linear viscoelastic properties. Neither strain
nor strain rate is small, and Boltzmann superposition
is no longer valid. The chains are displaced significantly from their equilibrium, random coil conformations. The organization of the chains - the physical structure of the liquid, so to speak - is altered
by the flow. The effects of finite strain appear,
much as they do in deformed polymer networks.

If a liquid is sheared at a constant shear rate
$\dot{\gamma}$, the stress will eventually reach a steady state.
In the parallel plate illustration in Fig. 12, the
upper plate moves at constant velocity V in
direction 1, the constant shear force F acts in
direction 1 on a face of the liquid normal to direction 2. The deformation here is uniform: the components of fluid velocity depend on shear rate, $\dot{\gamma} =$
V/H_2 and are proportional to the distance x_2 measured
from the fixed plate. The forces acting on each
element of the liquid are the same everywhere.

Suppose we could "cut out" a tiny cubical element of the liquid and examine the forces acting on
it at some moment. In a Newtonian liquid the components of force normal to the faces would all be the
same and would depend only on the local pressure.
There would also be a shear component in the force
acting on certain faces. Those components would also
be equal and directly proportional to the shear rate.
The situation is changed in an important way for a
viscoelastic liquid. The normal components of force
in the three directions are not always equal. The
differences depend on shear rate. The shear components, though equal, are not always directly proportional to shear rate.

The components of stress are simply the respective force components divided by the area of the
face. The stress is nothing more than the set of
these force/area components. Thus, aside from pressure, the stress in a viscoelastic liquid, at steady
state in simple shear flow, is completely defined by
three functions of shear rate, the shear stress function and two normal stress difference functions. The
labelling of stress components (the first subscript
denotes the face acted upon, the second denotes the
direction of the force component) should be clear
from the diagram. The shear stress is $\sigma(\dot{\gamma})$, the
first and second normal stress differences are $N_1(\dot{\gamma})$
and $N_2(\dot{\gamma})$. Each is zero for the liquid at equilibrium. The shear stress is proportional to shear
rate at very low shear rates; the normal stress
functions are proportional to the square of shear
rate. The steady state viscosity $\eta(\dot{\gamma})$ starts at η_o
and decreases with increasing shear rate. The first
normal stress coefficient $\Theta_1(\dot{\gamma})$ starts at $2 J_e^o \eta_o^2$ and

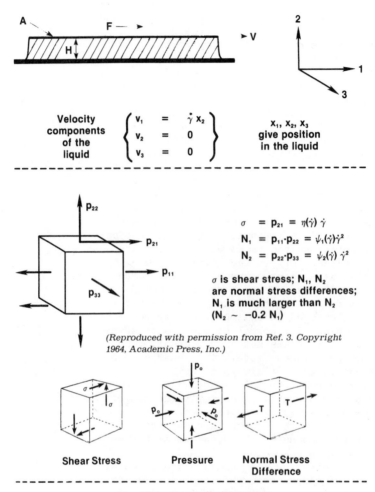

Velocity components of the liquid

$$
\left\{
\begin{array}{lcl}
v_1 & = & \dot{\gamma}\, x_2 \\
v_2 & = & 0 \\
v_3 & = & 0
\end{array}
\right\}
$$

x_1, x_2, x_3 give position in the liquid

$$\sigma = p_{21} = \eta(\dot{\gamma})\,\dot{\gamma}$$
$$N_1 = p_{11}\text{-}p_{22} = \psi_1(\dot{\gamma})\dot{\gamma}^2$$
$$N_2 = p_{22}\text{-}p_{33} = \psi_2(\dot{\gamma})\,\dot{\gamma}^2$$

σ is shear stress; N_1, N_2 are normal stress differences; N_1 is much larger than N_2 ($N_2 \sim -0.2\, N_1$)

(Reproduced with permission from Ref. 3. Copyright 1964, Academic Press, Inc.)

Shear Stress   **Pressure**   **Normal Stress Difference**

At sufficiently small shear rates:

$$\sigma = \eta_0\, \dot{\gamma}, \quad N_1 = 2\, J_e^0 \eta_0^2\, \dot{\gamma}^2$$

Figure 12. Steady-state shear flows.

also decreases with increasing shear rate. The
second normal stress difference is negative and
smaller than N_1 in magnitude ($-N_2/N_1$ is approximately
0.1-0.3 and roughly independent of shear rate).
Physically, the result is a net tension T, related
primarily to N_1, acting on all elements of the liquid
along the direction of flow. Viscosity-shear rate
behavior is relatively easy to measure (see below).
Measurement of first normal stress difference is more
difficult, especially at high shear rates, and data
on N_2 are very scarce indeed.

Figure 13 shows a diagram of the cone and plate
rheometer, a device which can be used to measure both
$\eta(\dot\gamma)$ and $N_1(\dot\gamma)$ at relatively low shear rates. The
liquid is placed in the gap, one of the fixtures is
held fixed , and the other is rotated at constant
speed $\dot\phi$. The steady state torque M and a thrusting
force F in the axial direction are measured. If the
gap angle is small, the shear rate is $\dot\gamma = \dot\phi/\alpha$ and is
the same everywhere in the liquid. The torque and
radius R give the shear stress; the axial force and
radius give the first normal stress difference. The
connection between axial force and N_1 is not too
difficult to understand. Tensions along the lines of
flow (see Fig. 12, bottom) cause the outer layers of
liquid to squeeze inward upon the inner layers, resulting in a pressure on the cone and plate which
builds up from near zero at the outer edge to a maximum at the center. The total axial force is just the
sum of contributions from this pressure. Incidental-
ly, this same geometry is commonly used to measure
the dynamic moduli. In this case a sinusoidal deformation of small amplitude is imposed and the in-phase
and out-of-phase components of the torque are measured. The axial force is small in the linear viscoelastic (small amplitude) experiments.

Data on σ and N_1 are shown at the bottom of Figure 13 for concentrated solutions of polyisoprene. At
low shear rates σ is much larger than N_1, but σ grows
as $\dot\gamma$ and N_1 as $\dot\gamma^2$. The curves cross and N_1 is larger
at high shear rates. Near the crossing point ($\sigma \sim$
N_1) σ begins to depart from proportionality to $\dot\gamma$,
i.e., the viscosity begins to decrease from η_o. That
qualitative characteristic appears to be quite general. The onset of non-Newtonian viscosity behavior
occurs near the shear rate where σ and N_1 become
equal, and N_1 becomes increasingly large compared to
σ at higher shear rates. The last figure shows the
data plotted as viscosity and normal stress difference divided by the square of shear stress vs shear
rate. In many cases N_1/σ^2 is fairly constant even
well beyond the knee of the viscosity curve, a property which is sometimes useful for extrapolation
purposes.

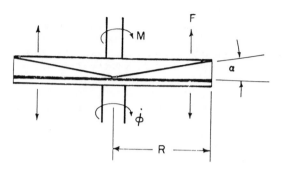

$$\dot{\gamma} \;=\; \dot{\phi}/\alpha \qquad \text{(small gap angle } \alpha\text{)}$$

$$\sigma \;=\; \frac{3M}{2\pi R^2} \qquad \text{(torque measurement)}$$

$$N_1 \;=\; \frac{2F}{\pi R^2} \qquad \text{(axial force measurement)}$$

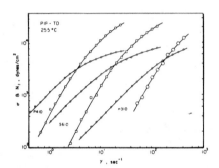

(Reproduced with permission from Ref. 4. Copyright 1975, Syracuse University Press.)

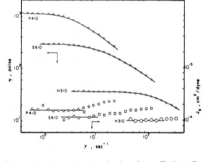

(Reproduced with permission from Ref. 4. Copyright 1975, Syracuse University Press.)

$$N_1 \;=\; 2J_e^{\circ}(\sigma_0\gamma)^2 \;=\; 2\,J_e^{\circ}\sigma^2 \text{ (small } \dot{\gamma}\text{)}$$

**Actually, $N_1/2\sigma^2 = J_s(\dot{\gamma})$ is "fairly constant"
even at higher shear rates (useful for extrapolation)**

Shear stress and first normal stress data for 10% solutions of polyisoprene, left, and viscosity η and normal stress compliance J_S as functions of shear rate for 10% polyisoprene solutions, right. Shear stress is indicated by small circles in each case. Normal stress is indicated by large symbols: linear P410 (◯), four-arm star S610 (◻), and six-arm star H310 (◯). *(Reproduced from Ref. 4. Copyright 1976, American Chemical Society.)*

Figure 13. Cone/plate rheometry: shear and normal stresses.

The use of cone and plate rheometers is limited
to relatively low shear rates by the onset of flow
instabilities, typically occurring not far beyond $\sigma =$
N_1 in polymer melts. A capillary rheometer is
sketched in Fig. 14. Operation at much higher shear
rates is possible in this instrument; viscosity vs.
shear rate can be determined, but not N_1. The viscosity for Newtonian liquids can be calculated from
pressure drop and flow rate data with Eq. 1. That
equation is not correct for non-Newtonian liquids
because it assumes a constant viscosity. Unlike cone
and plate flow, the shear rate in capillaries varies
with the position of the liquid element, in this
case with its distance from the centerline, thus producing different viscosities at different distances
from the centerline. It turns out, however, that
shear stress at the capillary wall can be calculated
from the pressure drop (Eq. 2) for any liquid, and,
remarkably, that shear rates at the wall can be
evaluated from Q vs ΔP data (Eq. 3), also for any
liquid. Thus, viscosity-shear rate data can be
obtained even for non-Newtonian liquids. Capillary
rheometry for viscoelastic liquids is more complicated than indicated here, however. There are
entrance effects and other factors to be accounted
for which we have no time to cover here.

Results for a commercial polystyrene sample are
shown in the bottom of Fig. 14. Cone and plate measurements cover the low shear rate range and capillary measurements the high shear rate range. The instruments provide complimentary information on the
viscosity behavior.

The sketch at the top of Fig. 14 also illustrates the phenomenon of die swell by which the extrudate rearranges to a larger diameter as it leaves the
capillary. The tension along the lines of flow (Fig.
12, bottom) draw the extrudate back when the confine-
ment by the capillary walls is no longer acting. The
liquid recoils. The swell ratio D/D_0 increases with
shear rate (or volumetric flow rate $\dot{Q}$), and, as shown
in Fig. 15, it also depends on the length/diameter
ratio of the capillary. The length dependence reflects the general memory effect in polymer melts:
time and deformation eventually rid the response of
contributions before the capillary entrance. We will
return to die swell later to discuss its nature
for the case of very long capillaries.

Time-temperature superposition works for non-
linear viscoelastic properties. Indeed, the temperature shift factors are indistinguishable from those
obtained in linear viscoelastic measurements. Stress
plays the role of modulus, and shear rate plays the
role of frequency. Curves of shear stress vs shear
rate are shifted by temperature along the shear rate

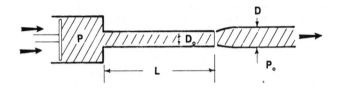

Q = volumetric flow rate (cm³/sec)

ΔP = P-P$_o$ = pressure drop (dyne/cm²)

D/D$_o$ = diameter swell ratio

Newtonian Liquid:

$$\eta = \frac{\pi D_o^4}{128L} \frac{\Delta P}{Q} \quad \text{(large L/D}_o\text{)} \tag{1}$$

Polymer Liquids (viscosity varies with shear rate):

$$\sigma_w = \text{shear stress at the wall} = \frac{D_o}{4L} \Delta P \tag{2}$$

$$\dot{\gamma}_w = \text{shear rate at the wall} = \frac{8Q}{\pi D_o^3} \left[3 + \frac{d \log Q}{d \log \Delta P} \right] \tag{3}$$

$$\eta(\dot{\gamma}_w) = \sigma_w/\dot{\gamma}_w \tag{4}$$

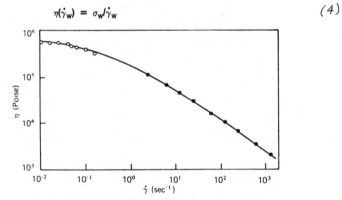

(Reproduced with permission from Ref. 4. Copyright 1975, Syracuse University Press.)

Figure 14. Capillary rheometry; viscosity and swell ratio.

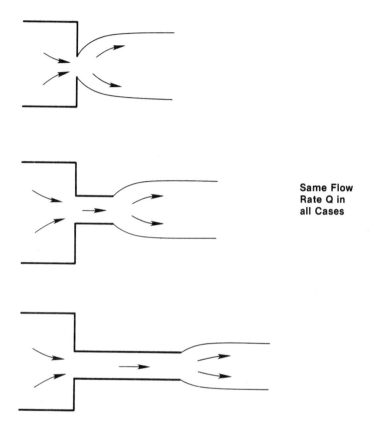

**Same Flow
Rate Q in
all Cases**

**Swell ratio varies with length/diameter
ratio of the capillary**

Figure 15. Memory effects.

axis without change of shape. The same data, plotted as viscosity-shear rate curves, shift by very nearly equal amounts along each axis, i.e., along a line of slope = -1 on log-log plots (Fig. 16, top). The next figure shows the effect of shifts proportional to η_o along both axes (along the line of slope = -1). The third figure shows that superposition can be achieved by allowing arbitrary shifts, represented by τ_o is this case. The difference between the last two figures has to do with the effect of an unusual temperature dependence in J_e^o for this polymer. It turns out that the values of τ_o are proportional to the product $\eta_o J_e^o$, a general relationship which will be discussed later. Another useful result, temperature independence of the relationship between σ and N_1 for a given polymer sample, follows from the temperature superposition principle as stated above. Both depend on shear rate, and thus shift with temperature, but when $\dot{\gamma}$ is eliminated to obtain $N_1 = f(\sigma)$, the result is quite insensitive to temperature.

Figure 17 at the top shows swell ratios (long capillary values) for a commercial polystyrene in relation to its viscosity-shear rate behavior. At low shear rates the viscosity levels off at η_o. The normal stress differences are small in this region, as discussed earlier, and D/D_o is about 1.1, the value for Newtonian liquids. The swell ratio begins to rise with the onset of shear rate dependence in the viscosity. Normal stress and shear stress have comparable magnitudes there, and beyond that region die swell increases steadily. Theories relating swell ratio and normal stress have had some mild success. Comparisons with data are shown in the bottom figure. Although testing is hampered by the lack of reliable normal stress data in the high shear rate region, Tanner's equation seems to be about the best of the lot, based on comparisons with data on several polymers. The phenomenon itself is complicated. The calculation of swell ratio even in Newtonian liquids is difficult.

The effect of temperature on die swell is shown in Fig. 18. At each shear rate the swell ratio decreases with increasing temperature. However, the data at all temperatures superpose rather well when plotted as a function of shear stress. That result turns out to be quite general and of course quite useful for extrapolation purposes. It is a natural consequence of the temperature invariance of the relationship between σ and N_1 and the idea, embodied by Tanner's equation (Fig. 17, bottom), that D/D_o depends on nothing more than the values of σ and N_1 in the capillary.

$$\eta\,(\dot\gamma)\ =\ \mathbf{a_T}\,\eta_r\,(\dot\gamma\;\mathbf{a_T})$$

or

$$\eta(\dot\gamma)\ =\ \sigma_o\,\eta_r\,(\dot\gamma\;\eta_o)/(\gamma_o)_r\ \textbf{(since}\ \eta_o\mathbf{(T)}\ \alpha\ \mathbf{a_T)}$$

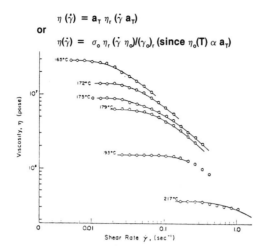

Viscosity vs. shear rate at various temperatures for linear polystyrene PC-4.

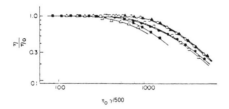

Reduced viscosity vs. reduced shear rate for linear polystyrene PC-4. Key: (▲), 165 °C; (○), 172 °C; (●), 175 °C; (□), 179 °C; (■), 193 °C; and (△), 217 °C.

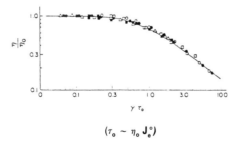

$$(\tau_o\ \sim\ \eta_o\,\mathbf{J_e^o})$$

Figure 16. Time-temperature superposition in flow properties. (Reproduced with permission from Ref. 6. Copyright 1974, John Wiley & Sons, Inc.)

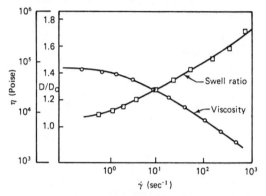

Comparison of viscosity and die-swell dependences on shear rate for a commercial polystyrene ($\overline{M}_w$ = 220,000). The temperature is 180°C.

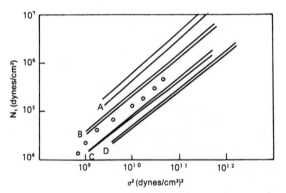

Normal stress difference as a function of shear stress for a commercial polystyrene ($\overline{M}_w$ = 220,000). The points are data measured in a plate-cone instrument. The lines are values calculated from die-swell behavior according to several theories: Curve A - Graessley, Glasscock, and Crawley; Curve B - Tanner; Curve C Mori, Curve D - Bagley and Duffey.

$$(N_1/2\sigma)^2 = 2\left[\left(\frac{D}{D_o} - 0.1\right)^6 - 1\right] \quad \text{(Tanner)}$$

Figure 17. Swell ratio in normal stress relationships. (Reproduced from Ref. 4. Copyright 1976, American Chemical Society.)

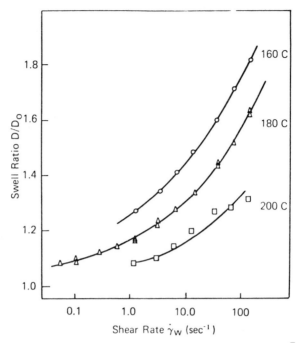

Die swell as a function of shear rate for a commercial polystyrene ($\bar{M}_w$ = 220,000). Data was gathered on nonannealed extrudates; D_0 = 0.070'', L/D_0 ranged from 27 to 56.

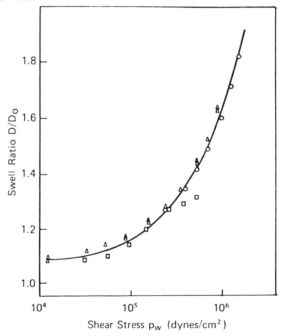

Die swell as a function of shear stress at the capillary wall.

Figure 18. Temperature effects on swell ratio. (Reproduced from Ref. 4. Copyright 1976, American Chemical Society.)

EFFECTS OF
MOLECULAR STRUCTURE

The importance of the terminal region in viscoelastic response related to flow behavior and its strong dependence on chain length, distribution and branching has already been discussed. We will now consider molecular effects in more detail. It is important to bear in mind the difference between local chain structure which, aside from effects induced by a previous flow history, controls the physical properties of polymers in the solid state, and large scale chain structure, which controls the viscoelastic properties of polymers in the liquid state (Fig. 19). Commercial polymers are typically very heterogeneous in large scale structure. Commercial polyolefins, containing appreciable numbers of molecules distributed over four or more orders of magnitude in chain length, are rather extreme examples, but in general the distributions are quite broad. Fortunately, the effect of molecular weight and molecular weight distribution appears to be essentially universal - they affect viscoelasticity according to general laws. These laws are independent of the local chain structure: aside from parameters which depend on the species, they are the same for all flexible chain polymers.

Figure 20 shows the molecular weight distribution for another commercial polymer, in this case, polyvinyl chloride. The weight distribution $W(M)dM$ is the fractional weight of the sample contributed by chains with molecular weight in the range, M, $M + dM$. The distribution is usually characterized by its averages, $\bar{M}_n$, $\bar{M}_w$, $\bar{M}_z$ and $\bar{M}_{z+1}$, defined as ratios of successively higher moments of the distribution. The values for this sample, not a particularly broad one compared with the polyolefins, are as shown, They were calculated from dilute solution measurements with a calibrated gel permeation chromatograph. The values of $\bar{M}_z$, and especially of $\bar{M}_{z+1}$, are only estimates; the results for those averages are invariably exceedingly sensitive to GPC baseline determinations. It turns out that rheological properties are far more sensitive to distribution, and particularly to the high molecular weight end of the distribution, than dilute solution methods. It is not unusual to find two polymer samples with GPC traces that are the same within the limits of reproducibility, and yet which differ significantly in their rheological properties. Obviously, it is essential to understand how molecular structure affects rheological properties and to use samples of well defined structures to explore those effects. Sensitivity of high molecular weight tails is, in fact, to be expected from the very simplest of molecular models. The discussion begins with a consideration of narrow distribution polymers,

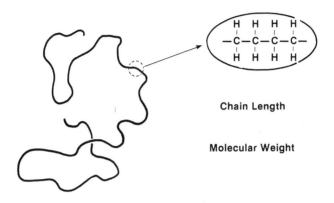

Chain Length

Molecular Weight

MOLECULAR WEIGHT DISTRIBUTION

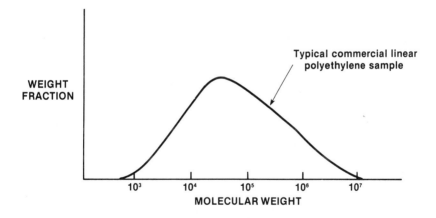

Figure 19. *Local and large-scale structure and molecular weight distribution.*

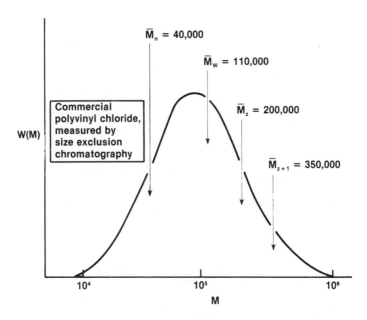

W(M) dM = fractional weight of sample contributed by
by chains with molecular weight in the
range M, M + dM

$$\overline{M}_n = \frac{\int W(M)\,dM}{\int \frac{1}{M} W(M)\,dM} \qquad \text{number-average molecular weight}$$

$$\overline{M}_w = \frac{\int M W(M)\,dM}{\int W(M)\,dM} \qquad \text{weight-average molecular weight}$$

$$\overline{M}_z = \frac{\int M^2 W(M)\,dM}{\int M W(M)\,dM} \qquad z\text{—average molecular weight}$$

$$\overline{M}_{z+1} = \frac{\int M^3 W(M)\,dM}{\int M^2 W(M)\,dM} \qquad z+1\text{—average molecular weight}$$

Figure 20. Molecular weight distribution and averages.

then considers polydispersity contributions, and finally describes some of the effects of long chain
branching.

As before, we consider flexible chain molecules
which are random coils at equilibrium. Random coils
with Gaussian statistics is an excellent approximation in polymer melts and concentrated solutions, as
discussed in the chapter on networks by Professor
Mark. Thus, the mean-square end-to-end distance
$<R^2>$ is proportional to chain length, and the radius
of gyration S^2 is related to $<R^2>$ by the formula S^2
$= <R^2>/6$ for linear chains.

Deformation distorts the distribution of conformations - the chains are carried continually into new
conformations by the frictional forces from the relative motions of their surroundings - and Brownian motion tends to restore equilibrium. The competition of
these two effects determines the average distortion
at any moment, and thus the stress.

The molecular model of Rouse (Fig. 21) represents the effects of flexible polymer chains moving
independently in a viscous medium. The motion of the
medium exerts a frictional force (the viscous force)
on each chain unit, governed by the monomeric friction coefficient ζ_o. Adjacent units along the chain
exert forces on one another, the connector force, to
preserve the integrity of the chain. Brownian motion
exerts a fluctuating random force (the osmotic
force). The real local connector forces are of
course very complicated, but, since we are primarily
interested in the large scale (slow) chain motions,
we can simplify locally without harm by "lumping"
the friction into widely spaced beads connected by
linear springs (the spring force). The spring constants are chosen to give the real chains and the
Rouse chains the same distribution of large scale
conformations at equilibrium: $<R^2>$ is the same for
both. The balance of forces on each bead leads to a
differential equation for Ψ, the conformational distribution function produced by the flow, and finally
to an expression for the stress, assuming the chains
contribute independently.

The Rouse model results for shear flow properties at steady state in monodisperse polymer liquids
are given in Fig. 22. The viscosity is predicted to
be directly proportional to molecular weight. The
only unknown quantity in the expression for η_o is the
monomeric friction coefficient ζ_o. That parameter
contains the temperature dependence of viscosity;
its value has been estimated in a few polymers from
the diffusion coefficient of monomer-size molecules
in the melt. The expression for recoverable compliance contains no unknown parameters. Its temperature
dependence is very weak, going only as $(\rho T)^{-1}$, a

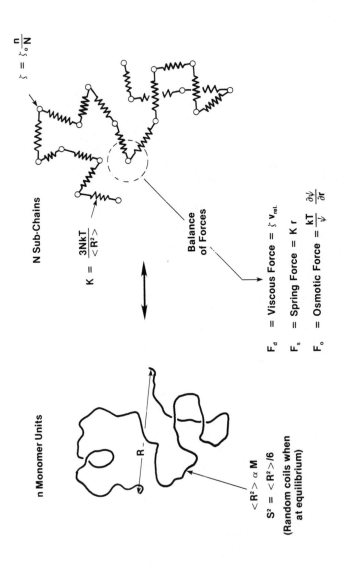

Figure 21. Rouse molecular model.

$$\eta_o = \left[\frac{\zeta_o N_a c}{6 \, m_o} \right] S^2 \qquad (= f(T) \, M \text{ for undiluted melts})$$

ζ_o = friction coefficient per monomer unit

N_a = Avogadro's number

m_o = molecular weight per monomer unit (M/n)

c   = polymer concentration (gm/cm^3)

c $\approx$ $\phi\varrho$, where ϕ is volume fraction of polymer for solutions, and ϱ is the polymer melt density

S^2 = equilibrium mean-square radius of gyration

S^2 = $<R^2>/6$

M   = molecular weight of the polymer

$$J_e^o = \frac{2}{5} \frac{M}{cRT} \quad (= \frac{2}{5} \frac{M}{\varrho RT} \quad \text{for undiluted melts})$$

$\eta(\dot\gamma) = \eta_o$ (all shear rates)

$$\tau_r = \frac{6}{\pi^2} \frac{\eta_o M}{cRT} \quad \text{(longest viscoelastic relaxation time)}$$

$$\tau_r = \frac{15}{\pi^2} \eta_o J_e^o, \text{ according to the Rouse model}$$

$\dot\gamma \, \tau_r << 1$ $\qquad\qquad\qquad\qquad\qquad\qquad$ $\dot\gamma \, \tau_r >> 1$

Figure 22. Rouse model predictions for linear chains.

result which is consistent with experimental behavior noted earlier, based on time-temperature superposition. The longest relaxation time τ_r is roughly $\eta_o J_e^O$. Viscosity is independent of shear rate, which is of course inconsistent with observations. As sketched at the bottom of Fig. 22, the product $\dot{\gamma}\tau_r$ governs the net distortion of conformation in shear flows at steady state.

How well does the Rouse theory work? Data on viscosity and recoverable compliance for polyisoprene as functions of chain length are shown in Fig. 23. The behavior for other species is very similar. Below a characteristic molecular weight M_c the viscosity prediction is generally rather good: η_o is proportional to M, and the magnitudes (using ζ_o from diffusion coefficient data) are rather near the experimental values. Above M_c the viscosity varies with a much larger power of molecular weight. (An exponent of 3.4-3.6 is typical). Below a somewhat larger characteristic value, M_c', the Rouse model prediction for J_e^O (shown by the dashed line) is not at all unreasonable. Above M_c', however, the recoverable compliance becomes independent of molecular weight. For undiluted polyisoprene $M_c = 10,000$ and $M_c' = 40,000$; the pattern is the same in other polymer species and also in concentrated solutions. Only the characteristic molecular weights are different.

There is also a characteristic molecular weight associated with the plateau modulus. We discussed the plateau modulus earlier, noting that a liquid of long chains acts at intermediate times and frequencies like a network. In the theory of rubber elasticity the equilibrium shear modulus is related to the concentration of network strands by the equation at the top of Fig. 24. That equation is used to calculate M_e, the entanglement molecular weight, from G_N^O.

Values of M_e, M_c and M' for several polymer species are given in the table at the bottom of Fig. 24. Note that these are relatively small compared with typical molecular weights for commercial polymers of the same species. Entanglement effects will certainly dominate the flow behavior in those systems. Although the values of M_e, M_c and M' vary with species, they are roughly in the ratios 1: 2:6 for a given species.

All three characteristic molecular weights increase as the polymer is diluted, and they seem to obey the same general dilution law. Thus, G_N^O is accurately proportional to a power of polymer volume fraction in concentrated solutions (Fig. 25). The exponent appears to be slightly larger than two and may vary slightly with the polymer species. Accordingly, M_e varies roughly as the reciprocal of the volume fraction. The same law, independent of polymer species, diluent species and temperature, applies

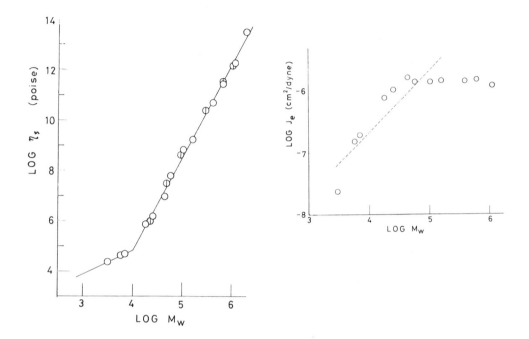

Figure 23. Left, the logarithmic plot of η_ζ at constant friction factor against M_w for poly(cis-isoprene). The two straight lines have the slopes of 1.0 and 3.7. Right, the logarithmic plot of J_e at -30 °C against M_w for poly(cis-isoprene). The dotted line represents the Rouse prediction. (Reproduced from Ref. 7. Copyright 1972, American Chemical Society.)

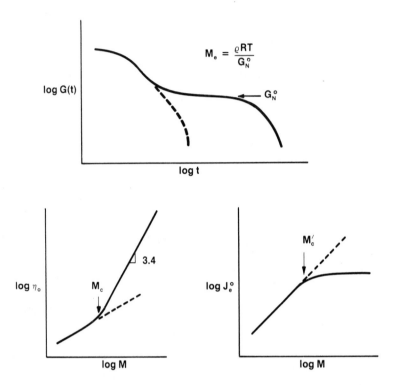

CHARACTERISTIC MOLECULAR WEIGHTS FOR UNDILUTED LINEAR POLYMERS

	M_e	M_c	M_c'
Polystyrene	19,000	36,000	130,000
Poly (α-methyl styrene)	13,500	28,000	104,000
1,4 Polybutadiene	1,700	5,000	11,900
Poly (vinyl acetate)	6,900	24,500	86,000
Poly (dimethyl siloxane)	8,100	24,400	56,000
Polyethylene	1,250	3,800	14,400
cis-Polyisoprene	6,300	10,000	28,000
Poly (methyl methacrylate)	5900 (10,000)	27,500	>150,000
Polyisobutylene	8,900	15,200	—

Figure 24. Characteristic molecular weights.

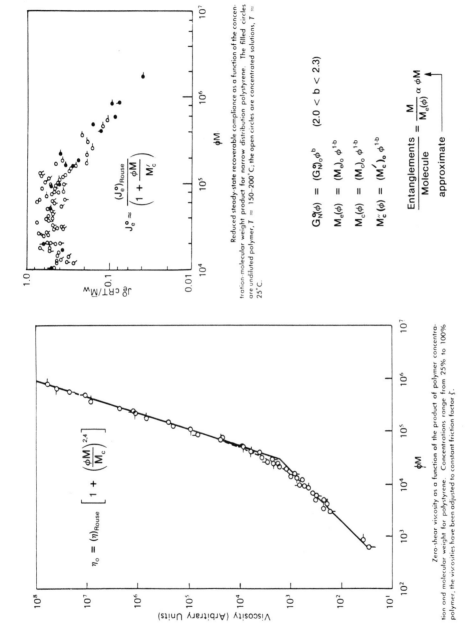

$$G_N^0(\phi) = (G_N^0)_o \phi^b \qquad (2.0 < b < 2.3)$$

$$M_e(\phi) = (M_e)_o\, \phi^{1-b}$$

$$M_c(\phi) = (M_c)_o\, \phi^{1-b}$$

$$M_c'(\phi) = (M_c')_o\, \phi^{1-b}$$

$$\frac{Entanglements}{Molecule} = \frac{M}{M_e(\phi)} \propto \phi M \quad \longleftarrow$$
approximate

Reduced steady-state recoverable compliance as a function of the concentration-molecular weight product for narrow distribution polystyrene. The filled circles are undiluted polymer, $T = 150–200°C$; the open circles are concentrated solutions, $T = 25°C$.

$$J_e^0 \approx \frac{(J_e^0)_{Rouse}}{\left(1 + \dfrac{\phi M}{M_c'}\right)}$$

$$\eta_o = (\eta)_{Rouse}\left[1 + \left(\frac{\phi M}{M_c}\right)^{2.4}\right]$$

Zero-shear viscosity as a function of the product of polymer concentration and molecular weight for polystyrene. Concentrations range from 25% to 100% polymer, the viscosities have been adjusted to constant friction factor ζ.

Figure 25. Effects of concentration on entanglement. (Reproduced from Ref. 4. Copyright 1976, American Chemical Society.)

to M_c and M_c' as well. For random mixing, the density
of polymer-polymer contacts goes as ϕ^2, and the
effective number of entanglements, though obviously
much smaller, varies in about the same way as the number of contacts. From this standpoint, the number of
entanglements per chain should be proportional to the
product ϕM.

Viscosity and recoverable compliance for melts
and concentrated solutions of narrow distribution,
linear polystyrenes are shown in Fig. 25. Arbitrary
shifts along the viscosity axis have been applied for
different concentrations and temperatures, solvent
species and concentration. Recoverable compliance
has been plotted in the reduced form $J_e^o cRT/M$, which
would be 0.4 for Rouse chains. The correlating variable in both cases is ϕM, which is proportional to
the number of entanglements per molecule. The data
are fairly well represented by the equations on the
figures. It comes as no surprise that entanglement
interactions increase the viscosity. However, aside
from stating the experimental fact - that entanglements make the liquid less elastically compliant than
unencumbered Rouse chains in steady state flows - it
is not at all obvious why J_e^o should suddenly become
independent of molecular weight and inversely proportional to the square of polymer concentration.

How does polydispersity affect the values of η_o
and J_e^o? The Rouse model predicts that viscosity
should be proportional to weight-average molecular
weight (Fig. 26). In fact, it is found that $\bar{M}_w$ is an
excellent correlating variable for the viscosity of
mixtures of chain lengths even well into the entanglement region: $\eta = K \bar{M}_w^{3.5}$. The Rouse model predicts that J_e^o should be the product of J_{Rouse}, the
value for monodisperse chains of the same weight-
average molecular weight, and a polydispersity fac-
tor $\bar{M}_z \bar{M}_{z+1}/\bar{M}_w^2$. That form, with J_{Rouse} replaced by
a constant (the observed behavior for highly entan-
gled monodisperse chains) does a reasonable job of
describing J_e^o in entangled mixtures. Several other
rheological mixing laws have been proposed, but the
$\bar{M}_z \bar{M}_{z+1}/\bar{M}_w^2$ factor still stands up well in comparison
as shown by the figure at the bottom of Fig. 26.

These observations on mixtures do not mean, of
course, that the Rouse model is a physically correct
description of what is happening on the molecular
level. They simply indicate that its relative weigh-
ing of contributions from chains of different size is
not too far from reality. It seems remarkable that
one could mix two components of different molecular
weight and obtain a recoverable compliance at steady
state that is higher than either by a factor of ten
or more. Qualitatively, that result is reasonable,
however, when you consider the recovery experiment

Rouse model:

$$\eta_o = f(T)\,\overline{M}_w$$

$$J_e^\circ = \frac{2}{5}\,\frac{\overline{M}_w}{cRT}\left(\frac{\overline{M}_z\overline{M}_{z+1}}{\overline{M}_w^2}\right)$$

$$\tau_r(\overline{M}) = \frac{6}{\pi^2}\,\frac{\eta_o\overline{M}_w}{cRT}\left(\frac{\overline{M}}{\overline{M}_w}\right)$$

Observed:

η_o is effectively a function of $\overline{M}_w$ alone

J_e° is extremely sensitive to distribution breadth, and especially to high molecular weight tails. Also, $\overline{M}_z\overline{M}_{z+1}/\overline{M}_w^2$ does a reasonable job, but difficult to measure!

$$J_e^\circ \approx J_e^\circ(\overline{M}_w)\left[\frac{\overline{M}_z\overline{M}_{z+1}}{\overline{M}_w^2}\right]$$

Monodisperse chain value

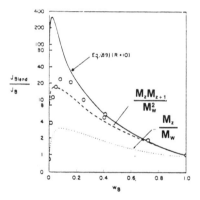

Experimental compliance data for mixtures of polydisperse silicones. The circles are the data for mixtures of $(M_w)_A = 58,500$ and $(M_w)_B = 596,000$. (Reproduced with permission from Ref. 8. Copyright 1971, American Institute of Physics.)

Figure 26. Effects of molecular weight distribution.

that defines J_e^o. The recoverable compliance J_e^o is simply the total recoil strain per unit stress. Longer chains have more frictional sites, and they also extend out farther in the surrounding flow field (S^2 increases with chain length for random coils). Thus, at steady state, the long chains support a disproportionate share of the stress in any mixture with shorter chains. For that reason, and also because they relax more slowly (see the equation for the τ_r (M) of mixtures in Fig. 26), the conformations of the longer chains are much more distorted at steady state than the shorter chains. Therefore, for a given shear stress, it is reasonable that the recoil strain from steady state should be larger for the mixture than for either of the pure components.

Turning now to viscosity-shear rate behavior (Fig. 27) we find a relatively clear experimental picture on molecular effects. Typical behavior is sketched at the top, showing the Newtonian region at low shear rates, going over smoothly to the power law region at high shear rates. The onset of shear rate dependence is located by a characteristic shear rate for the liquid, $\dot{\gamma}_o$. The viscosity curve, in reduced form, is found to depend only on molecular weight distribution for polymer melts and concentrated solutions. Thus, for example, data for narrow distribution samples of different polymer species, temperatures, concentrations and chain lengths can be superimposed simply by shifts in scale along the viscosity and shear rate axes. Moreover, the characteristic shear rate is related to η_o and J_e^o for the liquid. The product $\dot{\gamma}_o \eta_o J_e^o$ is a dimensionless constant of order unity. If, for example, $\dot{\gamma}_o$ is defined as the shear rate at which the viscosity has fallen to 0.8 η_o, the product $\dot{\gamma}_o \eta_o J_e^o$ is 0.6 ± 0.2. Data obtained for different molecular weights and concentrations are given in Fig. 27, and, at the top of Fig. 28, the universal reduced curve for narrow distribution polymers is shown. Data for polystyrene melts and solutions are used as examples, but the behavior is general.

The effect of molecular weight distribution is demonstrated in the second figure of Fig. 28. Data for two polystyrenes at 190°C are shown, one a commercial sample, the other a narrow distribution sample. The transformation from Newtonian to power-law behavior is broadened by polydispersity. The commercial (broad distribution) sample has a higher viscosity at low shear rates: it has a larger weight-average molecular weight so it has a larger value of η_o. However, non-Newtonian behavior begins at a much lower shear rate ($\dot{\gamma}_o$ is smaller for the commercial sample because both η_o and J_e^o are larger). The

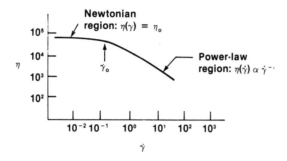

$\dot{\gamma}_o \, J_e^o \eta_o$ = universal constant of order unity

$$\frac{\eta(\dot{\gamma})}{\eta_o} = F\left(\frac{\dot{\gamma}}{\dot{\gamma}_o}\right) = F(\eta_o J_e^o \dot{\gamma})$$

Near-monodisperse polymers

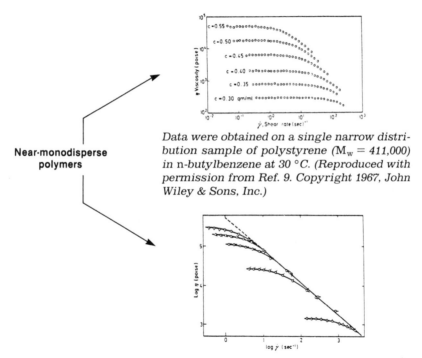

Data were obtained on a single narrow distribution sample of polystyrene ($M_w = 411,000$) in n-butylbenzene at 30 °C. (Reproduced with permission from Ref. 9. Copyright 1967, John Wiley & Sons, Inc.)

Weight-average molecular weights are ⌀, 48,500; ⌀, 117,000; ⌀, 179,000; ⌀ 217,000; and ⌀, 242,000. All data are reduced to 183 °C; the dashed line has a slope of -0.82. (Reproduced with permission from Ref. 10. Copyright 1966, Academic Press, Inc.)

Figure 27. Viscosity-shear rate behavior.

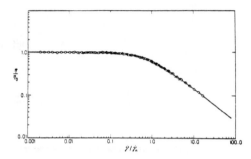

(Reproduced with permission from Ref. 11. Copyright 1974, Springer Verlag.)

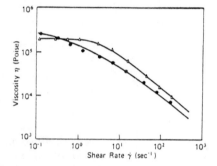

(Reproduced with permission from Ref. 4. Copyright 1975, Syracuse University Press.)

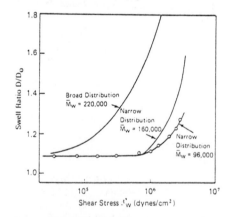

(Reproduced with permission from Ref. 4. Copyright 1975, Syracuse University Press.)

Figure 28. Top, universal curve for entangled near-monodis-perse polymers; middle, effect of molecular weight distribution on $\eta(\dot{\gamma})$; and bottom, effect of molecular weight distribution on swell ratio.

curves cross, and the broad distribution sample has a
lower viscosity at high shear rates. This combination of high viscosity at low rates and low viscosity at high rates is a desirable feature in certain
polymer processing operations - blow molding and film
blowing, for example.

There has been some success in predicted
viscosity-shear rate curves from the molecular weight
distribution. The curves drawn in the second figure
of Fig. 28 were calculated from GPC data, based on a
simple model which attributes the progressive reduction in viscosity with increasing shear rate to flow-
induced disentanglement of the chains.

Figure 28, bottom, demonstrates the extremely
important effect of distribution breadth on melt
elasticity. Increases in swell ratio D/D_0 appear at
much lower shear stresses in the broad distribution
sample, but the change in swell ratio with shear
stress is less abrupt. The sensitivity of die swell
to distribution breadth follows naturally from the
Tanner equation (Fig. 17, bottom) according to which
D/D_0 is a function of N_1/σ alone. Since $N_1 = 2J_e^o\sigma^2$
approximately (Fig. 13, bottom) and J_e^o increases
rapidly with distribution breadth (Fig. 26), the
ratio N_1/σ, and thus D/D_0, is expected to increase
with distribution breadth at constant shear stress.

All the earlier discussion about molecular
effects had to do with linear chain liquids. What
about the effect of branching? Broadly speaking, if
the branches are not too long the viscoelastic behavior is only slightly modified. However, if the
branches are long enough to be well entangled in
their own right, then their effects can be profound.
Non-linear structures in commercial polymers are
typically generated by some random branching chemistry during the polymerization. Random branching invariably yields a broad distribution of structures,
and it becomes extremely difficult to separate the
effects of distribution breadth alone from effects
due explicitly to the non-linear chain architecture.
Most of what is known about the latter has been established with model systems. Molecular stars (three
or more linear strands joined at a common junction)
of high uniform structure can be made by anionic
polymerization. Branch length and branch point functionality can be varied over wide ranges. Each molecule contains only one branch point (Fig. 29), so
some effects cannot be studied with stars. Only a
limited amount of data is available on regular combs
and on randomly branched chains, the latter prepared
by laborious fractionation of polydisperse samples.

The Rouse model has been extended to stars. The
equations for η_o and J_e^o are given in Figure 29. They
predict that the viscosity and recoverable compliance

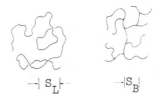

regular   regular   random
stars   combs   trees

At the same total molecular weight, the radius of gyration is smaller for a branched chain.

$\rightarrow|\ S_L\ |\leftarrow$ $\rightarrow|S_B|\leftarrow$

Rouse-Ham predictions:

$$\eta_o = f(T)\,gM$$

$$J_e^o = \frac{2}{5}\frac{g_2 M}{CRT}$$

$$g = \frac{3f-2}{f^2}$$

$$g_2 = \frac{15f-14}{(3f-2)^2}$$

$$g = \left(\frac{S_B^2}{S_L^2}\right)_M$$

For equal-arm stars with f branches

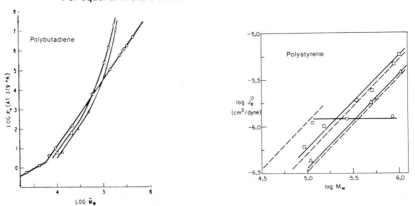

Left, dependence of Newtonian viscosity on molecular weight: ○, *linear;* □, *trichain; and* △, *tetrachain. (Reproduced from Ref. 12. Copyright 1965, John Wiley & Sons, Inc.) Right, zero-shear recoverable compliance vs. molecular weight:* ○, *linear polymer;* □, *four-arm stars;* △, *six-arm stars; and* ---, *Rouse-Ham equation. (Reproduced from Ref. 13. Copyright 1979, American Chemical Society.)*

Figure 29. Effect of long branches.

are both smaller for branched chains than linear
chains of the same molecular weight. The reduction
factors for random coils, g and g_2, can be calculated
from the structure. Expressions are given for stars
with f arms of equal length.

It is found that viscosities for branched and
linear polymers are approximately equal when compared
at the same radius of gyration, but only when the
branches are not too long. That dependence on size
alone persists well into the entanglement region.
The viscosity of both linear and branched versions
are increased by entanglement, but by about the same
factor. However, as the branches become longer
(typically, for arm molecular weights exceeding 2 to
4 times M_c), the branched polymer viscosity begins to
rise extremely rapidly (Fig. 29, bottom). The viscosity enhancement (relative to linear chains of the
same size) can easily reach factors of 100 or more.
Viscosity no longer has a simple power law dependence
on molecular weight. The enhancement of viscosity
appears, in fact, to be an exponential function of
the arm length. Enhancement decreases rapidly as the
polymer is diluted, and eventually the viscosity
returns to a dependence on molecular size alone.

In contrast, the recoverable compliance for
stars behaves in a remarkably simple way. The data
for melts and concentrated solutions conform to the
Rouse expression for stars at seemingly all arm
lengths. It rises in direct proportion to molecular
weight, and the numerical values are in good agreement with the Rouse model predictions for stars.
Unlike the linear chain behavior, J_e^o appears not to
level off to a constant value. Thus, values of J_e^o
for long arm stars may be quite large compared with
their linear chain counterparts.

What does this mean in regard to the other flow
properties? The effects on viscosity-shear rate
behavior are shown in Fig. 30. First, the earlier
comments on shear rate dependence of viscosity apply
equally well to stars. In reduced form ($\eta(\dot{\gamma})/\eta_o$ vs
$\dot{\gamma}/\dot{\gamma}_o$) the viscosity-shear rate curve for stars is
indistinguishable from data on linear chains with
narrow distributions. Perhaps in polydisperse
systems it now depends on the molecular size distribution, but that is uncertain. Second, the connection between characteristic shear rate $\dot{\gamma}_o$ and the
values of η_o and J_e^o is not changed by branching.
The product $\dot{\gamma}_o \eta_o J_e^o$ appears to be universal. Third,
the large values of J_e^o for long-arm stars mean that
$\dot{\gamma}_o$ is smaller than the value for a linear chain
liquid of the same viscosity. This means that the
viscosity at high shear rates is smaller for stars.
The viscosity-shear rate curves for linear and star
polymer liquids may cross, as illustrated in Fig. 30.

$\dot{\gamma}_0 J_e^\circ \eta_0$ = constant, which is the same constant for linear and branched polymers. Since J_e° is higher, $\dot{\gamma}_0$ is smaller for branched polymers (compared to linear polymers of the same η_0).

The viscosity-shear rate master curve appears to be independent of branching, depending only on molecular weight (or perhaps molecular size) distribution.

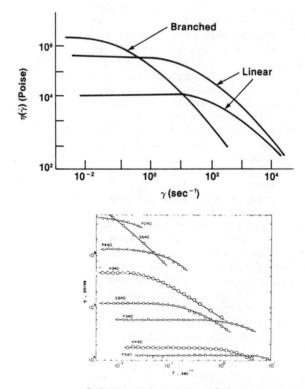

Solutions of various polyisoprenes. (Reproduced from Ref. 5. Copyright 1976, American Chemical Society.)

Figure 30. Viscosity-shear rate behavior in branched polymers.

Branched systems in general can reach domains of viscosity-shear rate behavior that are unattainable by
systems of linear chain. Unusual elastic effects are
presumably present as well, but that is still a relatively unexplored area.

We've now discussed some aspects of linear viscoelasticity and flow properties in flexible polymer
melts and their concentrated solutions. We've examined some effects caused by molecular weight distribution and long chain branching. We've seen that
many phenomena are universal, and we've talked somewhat vaguely about large scale chain motions and
chain entanglement, bringing in the Rouse model as a
kind of vehicle to help organize the facts. But -
what is really happening on the molecular level? Is
there some definite picture or testable hypothesis
about how the uncrossability of chains affects molecular motions, assuming, of course, that uncrossability is the key to understanding what we call
entanglement effects? Until a few years ago the
answers would have been somewhat unsatisfactory - a
beautiful set of general experimental laws but little
in the way of detailed fundamental understanding.
Recent developments appear to have changed that
situation, and I'd like to use the last few pages
to discuss them.

RECENT MOLECULAR
THEORIES OF
VISCOELASTICITY

Figure 31 illustrates the problem of individual
chain motion in an environment filled with chains.
In 1971 deGennes proposed that we think about the
uncrossability problem in the following way. Chains
at high concentration provide a kind of mutually
shared meshwork, with each chain lying along its own
tunnel through the mesh. The chain can't move side-
ways very far without crossing through other chains,
which is forbidden. It can, however, diffuse freely
along its own tunnel. He named that motion reptation
from its snake-like character, and he postulated that
reptation would provide the fastest path for rearrangement of the large scale chain conformation in linear chains. In 1975 he noted that a long branch in
the molecule would prevent reptation or at least
drastically retard it. He was actually treating the
movement of unattached molecules through a permanent
network in these cases, but he speculated that the
results might be valid even for a liquid, in which
all chains can move.

In 1978 Doi and Edwards developed a theory of
chain dynamics and viscoelastic behavior for linear
chain liquids based on the reptation idea. The "tube
model" in Fig. 32 attributes stress relaxation to the
diffusion of chains to new conformations by reptation
(illustrated in Fig. 33). Each molecule has the

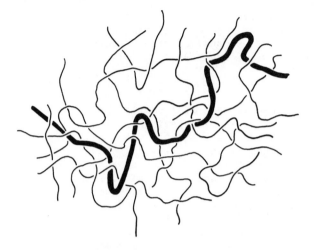

1 Damped network-like vibrations
2 Translation along the path (reptation)
3 Transverse displacements
   (a) Projection of loops
   (b) Release of constraints

- -

For pure reptation (de Gennes, Doi, Edwards):
1 Chains move like Rouse chains with constraints.
2 Constraints are represented by a tube.
3 The tube diameter corresponds to the mesh size.

Figure 31. Chain motions in concentrated systems. (Reproduced with permission from Ref. 14. Copyright 1982, Springer Verlag.)

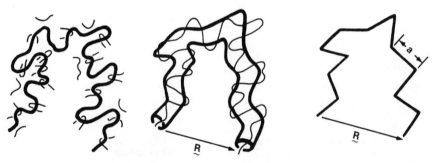

a = primitive path step length   L = primitive path length
N = number of steps along the path   L = Na ; $<R^2> = Na^2$

Figure 32. Doi–Edwards tube model. (Reproduced with permission from Ref. 14. Copyright 1982, Springer Verlag.)

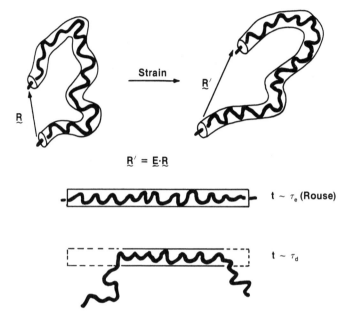

$$\underset{\sim}{R}' = \underset{\sim}{E} \cdot \underset{\sim}{R}$$

$t \sim \tau_e$ (Rouse)

$t \sim \tau_d$

Fraction of initial tube still occupied:

$$F(t) = \frac{8}{\pi^2} \sum_{\substack{odd \\ p}} \frac{1}{p^2} \exp\left[-\frac{p^2 t}{\tau_d}\right]$$

$$\tau_d = \frac{L^2}{\pi^2 D^*} \qquad D^* = \frac{kT}{\zeta_o n} \text{ (Rouse)}$$

$$\boxed{G(t) = G_N^o \, F(t)}$$

Figure 33. Relaxation after step strain.

dynamics of a Rouse chain, but now subject to the spatial constraints of the tube. That theory provided a basis for recent extensions to long-arm star molecules, where reptation is suppressed, and conformational rearrangement proceeds by the much slower process of "tube length" fluctuations. A third process has also been considered - constraint release - which takes into account the finite lifetime of the restrictions on lateral chain motion in liquids.

What has come from this so far is a predicted relationship between self-diffusion coefficient of chains and linear viscoelastic parameters which has been confirmed quantitatively (Fig. 34). The theory also predicts that J_e^o is independent of chain length for linear chain liquids (Fig. 35), a result that is consistent with experiment. The viscosity is predicted to vary as M^3 for linear chain liquids. That is not correct, but there is reason to believe that viscosity is sensitive to the contributions of all three processes - reptation, tube length fluctuations and constraint release - and that what is represented as $\eta_o \propto M^{3.4}$ in the observed range of chain lengths may not be a real power law, but rather the result of a very slow asymptotic approach to M^3 behavior for very long chains. For star polymers (Fig. 36) J_e^o is predicted to have a form which is indistinguishable from the equation for Rouse model stars. Also, the viscosity for stars is predicted to be an exponential function of arm length. Both results are consistent with the experimental observations.

In summary, the initial indications for this new approach appear to be favorable, and we may hope to see some real advances in molecular understanding as these ideas are developed more fully in the years ahead.

$$D = \frac{G_N^o}{135} \left(\frac{\varrho RT}{G_N^o}\right)^2 \frac{<R^2>}{M} \left(\frac{M_c}{\eta_o(M_c)}\right) \frac{1}{M^2}$$

$$\boxed{D = \frac{0.34}{M^2} \frac{cm^2}{sec} \text{ (predicted)}}$$

(Polyethylene melts at 176°C)

$$\boxed{D = \frac{0.26}{M^2} \frac{cm^2}{sec} \text{ (observed)}}$$

$3600 < M < 23{,}000$ in $\overline{M}_w = 160{,}000$ Matrix

Figure 34. Diffusion of polymers in the melt.

$$\eta_o = \frac{15}{4} \left(\frac{G_N^o}{CRT}\right)\left(\frac{\eta_o(M_c)}{M_c}\right)M^3 \qquad (M >> M_e)$$

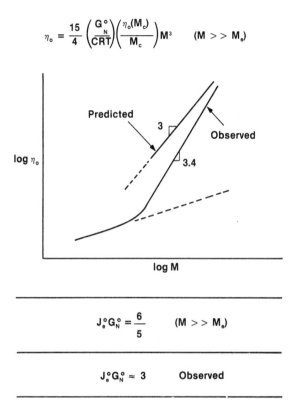

$$J_e^o G_N^o = \frac{6}{5} \qquad (M >> M_e)$$

$$J_e^o G_N^o \approx 3 \qquad \text{Observed}$$

Figure 35. Viscoelastic predictions.

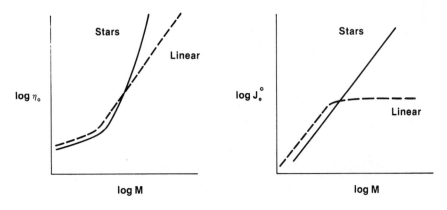

η_o is exponential in arm length

$J_e^o \approx$ Rouse model value (stars)

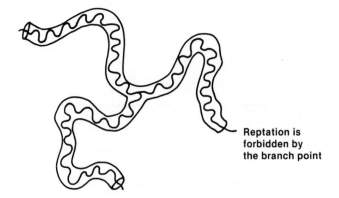

Reptation is
forbidden by
the branch point

Figure 36. Star-branched polymers.

LITERATURE CITED

1. Onogi, Shigeharu; Masuda, Toshiro; Kitagawa, Keishi Macromolecules 1970, 3 (2), 109.
2. Masuda, Toshiro; Kitagawa, Keishi; Inoue, Toshio; Onogi, Shigeharu Macromolecules 1970, 3 (2), 116.
3. Lodge, A. S. "Elastic Liquids"; Academic: New York, 1964.
4. "Characterization of Materials in Research"; Burke, J. J.; Weiss, V., Ed.; Syracuse Univ. Press: Syracuse, 1975.
5. Graessley, W. W.; Masuda, T.; Roovers, J. E. L.; Hadjichristidis, N. Macromolecules 1976, 9 (1), 127.
6. J. Polym. Sci., Polym. Phys. Ed. 1974, 12, 1771.
7. Nemoto, Norio; Odani, Hisashi; Kurata, Michio Macromolecules 1972, 5 (4), 531.
8. J. Chem. Phys. 1971, 54, 5143.
9. Trans. Soc. Rheol. 1967, 11, 267.
10. J. Colloid Sci. 1966, 22, 517.
11. Adv. Polym. Sci. 1974, 16, 1.
12. J. Polym. Sci., Part A 1965, 3, 105.
13. Graessley, W. W.; Roovers, J. Macromolecules 1979, 12 (5), 959.
14. Adv. Polym. Sci. 1982, 47, 67.
15. Klein, J. Nature (London) 1978, 271, 143.

ANNOTATED BIBLIOGRAPHY

1. J. D. Ferry, "Viscoelastic Properties of Polymers", 3rd Ed., Wiley, New York, 1980.

A general source of information on linear viscoelastic properties of polymers. Deals with experimental methods, molecular theories, analysis of data, and the effects of temperature, concentration and molecular structure in a wide range of polymer systems.

2. A. S. Lodge, "Elastic Liquids", Academic Press, New York, 1964.

An introduction to the physics and mathematics of polymer rheology. Develops the equations from first principles. A demanding but useful book for working, step by step, towards the background needed to understand the current rheological literature.

3. W. W. Graessley, "The Entanglement Concept in Polymer Rheology", Adv. Polymer Sci., 1974, 16, 1.

Emphasizes linear and non-linear viscoelastic behavior of linear polymers with narrow molecular weight distributions.

4. W. W. Graessley, Accts. of Chemical Research, 1977, 10, 332.

A brief review of the effects of branching on polymer rheology.

5. W. W. Graessley, "Entangled Linear, Branched and Network Polymer Systems", Adv. Polymer Sci., 1982, 47, 67.

A review of current molecular theories with references to reptation and the tube model.

6. W. W. Graessley, Chapter 15 in "Characterization of Materials in Research", Ed. J. J. Burke and V. Weiss, Syracuse University Press, Syracuse, 1975.

A brief summary of the relationships between molecular structure and the steady state flow properties, including die swell, of polymer melts.

7. K. Walters, "Rheometry", Chapman and Hall, London, 1975.

An excellent discussion of experimental methods from measuring the flow properties of polymer liquids.

8. R. B. Bird, R. C. Armstrong, O. Hassager "Dynamic Properties of Polymeric Liquids", Vol. 1 and 2 (C. F. Curtiss is a co-author of the second volume), Wiley, New York, 1977.

The first several chapters of Volume 1 are a superb introduction to the fluid mechanics of non-Newtonian liquids for people who already have some familiarity with Newtonian fluid mechanics. The second volume deals with molecular theories, but primarily those applicable to dilute solutions.

9. G. Astarita and G. Marrucci, "Principles of Non-Newtonian Fluid Mechanics", McGraw-Hill, Maidenhead, 1974.

There is no way to avoid the fact that polymer rheology is a branch of mechanics and requires some familarity with tensors to understand the description of both stress and deformation. That aspect is developed clearly in this book which also uses a terminology that is consistent with much of the current literature on the subject. Molecular aspects are not considered.

10. H. Janeschitz-Kriegl, "Polymer Melt Rheology and Flow Birefringence", Springer-Verlag, Heidelberg, 1983.

A useful discussion of current results and theories which emphasizes the use of optical measurements to obtain rheological information.

11. P. G. deGennes, "Scaling Concepts in Polymer Physics", Cornell University Press, Ithaca, 1979.

Structure and dynamics of polymer liquids from the physicist's point of view. Thought provoking and filled with nice insights on viscoelasticity and related topics.

4

The Crystalline State

Leo Mandelkern

The study of crystalline polymers closely parallels
the development of polymer science itself. To put
the subject in the perspective of the present time
there are certain areas which are well understood and
generally accepted. In contrast there are others
which are still under intensive study. The guiding
principles needed to understand the thermodynamics
of fusion, and crystallization kinetics are firmly
established by both theory and experiment. Modern
emphasis has therefore been directed to the influence
of the structure and morphology of a crystalline
polymer on its properties. The latter subject rests
very heavily on the first. We shall, therefore, be
considering both during the course of our discussion.
Since the thermodynamics and crystallization kinetics
are well documented in the literature, we shall be
content to briefly review these areas here and to
establish a base by developing the salient features.
Our major emphasis will be on the structure-property
relations which we shall discuss in some detail.
The major and underlying concern here is in the
development of basic principles, which will be
demonstrated with many examples. We want to emphasize the approach that is being taken. Once the
principles are understood they can be applied in the
resolution of a variety of problems. We begin by
considering the structure of individual polymer
molecules.

  Long chain molecules can exist in either of two
states which are characterized by the conformation
of the individual molecular chains and their

0851/84/0155$13.80/1
© 1984 American Chemical Society

organization relative to one another. The liquid state is the state of molecular disorder. In this state the individual chains adopt a statistical conformation, which is commonly called the random coil. The center of mass of the molecules are also randomly arranged relative to one another. All the thermodynamic and structural properties observed in this state are those which are commonly associated with a liquid, although in this case it is a very viscous one. The liquid state in polymers is also commonly called the amorphous state. It is this state which exhibits the characteristic long-range elasticity.

The crystalline or ordered state is one that is characterized by three-dimensional order over at least a portion of the chains. The ordered conformation may be fully extended or may represent one of many known helical structures. Irrespective of the details of the ordered chain structure, the molecules are organized into a regular three-dimensional array. The chain axes are aligned parallel to one another and the substituent groups are brought into regular register. Such ordered systems diffract x-rays in the conventional manner and display all the properties characteristic of the crystalline state. It can be stated as a general principle that all chain molecules which have a reasonable structural regularity will crystallize, under suitable conditions.

In contrast to the elasticity characteristic of the liquid state, the crystalline state is relatively inelastic and very rigid. For example, there is a difference of about five orders of magnitude between the modulus of elasticity in the two states. Major differences also exist in other properties, as for example, spectral and thermodynamic ones. Moreover, within the crystalline state it is also possible to change properties by control of structure. This ability to control properties turns out to be of major concern in the application and end use of this type polymeric system.

The conformational differences of the molecules in the two states is schematically illustrated in Figure 1. In the crystalline state the bonds adopt a set of successive preferred orientations; participation of the complete molecule in the ordering process is not required. In the liquid state the bond orientations are such that the chain adopts a statistical conformation.

Our primary interest is to learn how the properties of the crystalline state are influenced by the chemical nature of the repeating unit, the crystallite structure above the level of the unit cell, and the crystallite organization. One can be concerned with a number of properties, which range from what

appear to be simple thermodynamic ones, to rather
complex mechanical behavior and to the problems of
ultimate strength.

We initiate our discussion of the crystalline state
by outlining the basic foundations of the subject.
From the point of view of formal thermodynamics, it
has been established that the transformation from
one state to another can be properly treated as a
first-order phase transition in the classical sense.
The transformation is very similar to the fusion of
low molecular weight substances. A typical example
of the melting process for homopolymers is given in
Figure 2 for unfractionated and fractionated
samples of linear polyethylene. When carried out
carefully, the melting process is relatively sharp
and a well-defined melting temperature is clearly
seen to exist. It is easily seen in Figure IV.B.1
that for the molecular weight fraction fusion takes
place over a very narrow temperature interval. From
the specific volume-temperature plot the disappear-
ance of the last traces of crystallinity, which
defines the melting temperature can be clearly dis-
cerned in both examples.

The application of formal phase equilibrium
thermodynamics leads to an expression for the de-
pressions of the melting temperature by low molecular
weight diluent. This expression is given by Eq. (1)
of Figure 3. Here T_m^{0} is the equilibrium melting temperature; T_m the melting temperature corresponding to a volume fraction of diluent v_1; V_u/V_1 is
the ratio of the molar volume of the chain repeating
unit to that of the diluent, χ_1 is the polymer-
diluent thermodynamic interaction parameter; and ΔH_u
is the enthalpy of fusion per chain repeating unit of
the completely crystalline polymer. ΔH_u is characteristic of the chain repeating unit and does not
depend on the nature of the crystalline state.
Equation (1) is simply the adaptation to polymers of
the classical freezing point depression expression.
It has received widespread experimental verification
for many different polymers. The same value of ΔH_u
is obtained for a given polymer when studied with a
series of different diluents. Thus, by use of this
equation we can obtain ΔH_u for a given polymer.
Combined with the equilibrium melting temperature,
ΔS_u, the entropy of fusion per repeating unit is
obtained. These thermodynamic parameters for a
selected set of polymers are given in the table of
Table I. This table is not meant to be exhaus-
tive. The examples have been selected to illustrate
typical key situations.

THERMODYNAMICS
OF CRYSTALLIZATION

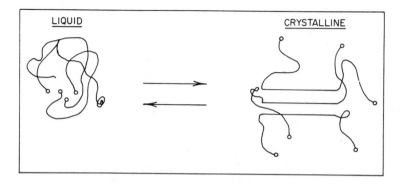

Figure 1. Schematic diagram illustrating conformational differences of chain molecules in liquid and crystalline state. Straight line represents ordered conformation. Details of interfacial structure are not being considered at this point.

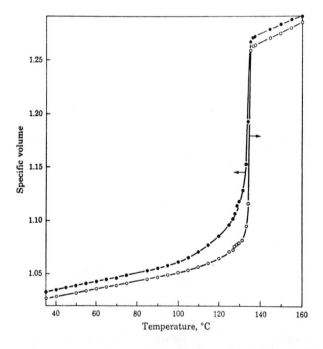

Figure 2. The melting of linear polyethylene: specific volume-temperature relations. Key: ●, unfractionated polymer; ○, fraction M = 32,000. (Reproduced from Ref. 1. Copyright 1961, American Chemical Society.)

$$(1) \quad \frac{1}{T_m} - \frac{1}{T_m^\circ} = \left(\frac{R}{\Delta H_u}\right)\left(\frac{V_u}{V_1}\right)\left[v_1 - \chi_1 v_1^2\right]$$

T_m° = equilibrium melting temperature of pure homopolymer

T_m = equilibrium melting temperature of polymer-diluent mixture

v_1 = volume fraction of diluent

V_1 = molar volume of diluent

V_u = molar volume of repeating unit

χ_1 = thermodynamic interaction parameter

ΔH_u = enthalpy of fusion per repeating unit

R = gas constant

Figure 3. Melting temperatures of polymer-diluent mixtures.

Table I. Thermodynamic Quantities Characterizing the Fusion
of Selected Polymers

Polymer	T_m^o °C[a]	ΔH_u[b] cal. mole^{-1}	ΔS_u cal. deg.$^{-1}$mol^{-1}
Polyethylene	146 ± 1	960	2.3
Polypropylene	200	1386	2.90
Poly(isoprene)1,4 cis	28	1050	3.46
Poly(isoprene)1,4 trans	74	3040	8.75
Poly(chloroprene)1,4 trans*	80	2000	5.68
Polystyrene-isotactic	243	2000	3.9
Polyoxymethylene	180	1590	3.5
Polyoxyethylene	80	1980	5.35
Polydecamethylene adipate	79.5	10200	29
Polydecamethylene sebacate	80	12000	34
Polydecamethylene terephthalate	138	11000	27
Polytetramethylene terephthalate	230	7600	15.1
Polydecamethylene sebacamide	216	8300	17
Polydecamethylene azelamide	214	8800	27
Cellulose tributyrate	207	3000	8.1

(a) Best estimate of equilibrium melting temperature
(b) ΔH_u determined from the depression of the melting temperature by
monomeric diluents
  * For highest melting polymorph
ΔH_u = enthalpy of fusion per repeating unit; ΔS_u = entropy of fusion
per repeating unit

For a more detailed discussion of thermodynamic parameters see Refs.
(2) and (3).

The data in this figure illustrate the guiding structural principles which determine the melting temperature. It is clear from these examples that there is no correlation between the melting temperature and the enthalpy of fusion. The values for ΔH_u fall into two classes. They are either of the order of just a few thousand calories per mole or about 10,000 cal. mole. For both the examples given here, and as is also more generally found, many high melting polymers possess low heats of fusion while conversely a large number of low melting polymers possess high values for the heat of fusion. Consequently, the entropy of fusion is a key factor in establishing the location of the melting temperature. A very striking causal relation can be developed between the entropy of fusion and the chain conformation in the completely molten state. Hence polymers commonly designated as elastomers have relatively low melting temperatures, and high entropies of fusion which reflect the compacted, highly flexible nature of the chain. So-called engineering plastics at the other extreme, have high melting temperatures and more extended chain structures, with the accompanying lower entropies of fusion. Cellulose derivatives are another case in point. As a class of polymers, they are characterized by high melting points and low heats of fusion. The low entropy of fusion must result from the highly extended nature of the chain.

The introduction of ring structures into a linear chain substantially raises the melting temperature, relative to the aliphatic chain, as would be expected from the decreased conformational entropy of the melt. Striking examples of this phenomenon are found in comparing the melting temperatures of the aliphatic and aromatic polyesters and polyamides.

Another example of the influence of the entropy of fusion is found in comparing aliphatic polyesters and polyamides. For corresponding type repeating units the melting temperatures of the polyamides are well-known to be substantially higher. Despite the hydrogen binding capacity of the polyamides, there is no substantive difference in the enthalpy of fusion of the two type chains. Hence the 150-200° difference in melting temperature must result from differences in the entropy of fusion.

From the few examples that have been described it should be apparent that as a general rule, the chain structure influences the melting temperature through its conformational properties and thus the entropy of fusion.

By applying classical phase equilibrium theory the melting point of a copolymer, relative to that of the parent homopolymer can be calculated. It is very important to note, and crucial to understand, that from the point of view of crystallization behavior, different type chemical repeating units, as well as structural irregularities, such as stereoirregularities, branch points or geometric irregularities, when incorporated into the chain all behave as copolymeric units. For the very common situation where these co-units or structural irregularities do not participate in the crystallization, Eq. (1), of Figure 4 is obtained. In this equation the quantity p represents the sequence propagation probability, i.e. the probability that in the copolymers a crystallizable unit is succeeded by another such unit. T_m° and ΔH_u are the same as already defined, T_m is the melting temperature of the copolymer, and X_A is the mole fraction of crystallizing units. We thus have the very interesting and most unique result that the melting temperature of a copolymer does not depend directly on its composition but instead on the nature of its sequence distribution.

For example, for a random type copolymer, where $p = X_A$ Eq. (2) results. For an ordered, or block, copolymer p is much greater than X_A and in some cases will even approach unity. For such copolymers there will at most be only a slight depression of the melting temperature from that of the corresponding homopolymer. On the other hand, for an alternating type copolymer, where p is much less than X_A there will be a rather drastic reduction in the melting temperature. From a theoretical point of view, we can expect copolymers which have exactly the same composition to have drastically different melting temperatures, depending on the sequence distribution. This conclusion is fulfilled experimentally as is illustrated in the following.

Figure 5 gives some typical examples of the melting temperature-composition relations for a set of random copolyesters and copolyamides. As is predicted by theory, there is a monotonic decrease in the melting temperature with increasing concentration of the non-crystallizing co-unit. The equivalent of a eutectic temperature is reached at an appropriate composition commensurate with the melting temperature of each component.

In Figure 6 the melting temperature-composition relations are given for blocked poly-(ethylene terephthalate) copolymers with different co-units. The data for several random copolymers are also included for reference. The difference between the melting temperature relations for the two types of copolymers are quite marked and are in

(1) $\quad \dfrac{1}{T_m} - \dfrac{1}{T_m^\circ} = \dfrac{-R}{\Delta H_u} \ln p$

For random copolymer $p \equiv X_A$

(2) $\quad \dfrac{1}{T_m} - \dfrac{1}{T_m^\circ} = \dfrac{-R}{\Delta H_u} \ln X_A$

For block copolymer $p \gg X_A \rightarrow 1$

For alternating copolymer $p \ll X_A$

T_m° = equilibrium melting temperature of homopolymer

T_m = equilibrium melting temperature of copolymer

p = sequence propagation probability

X_A = mole fraction of crystallizing A units

ΔH_u = enthalpy of fusion per repeating A unit

R = gas constant

Figure 4. Melting temperature of copolymers.

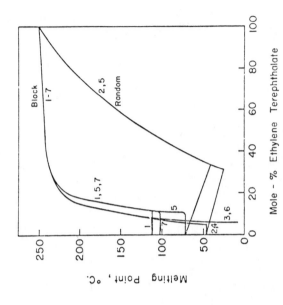

Figure 6. Melting temperature-composition relations for block copolymers of polyethylene terephthalate with (1) ethylene succinate; (2) ethylene adipate; (3) diethylene adipate; (4) ethylene azelate; (5) ethylene sebacate; (6) ethylene phthalate; and (7) ethylene isophthalate. For comparative purposes data for random copolymers with ethylene adipate and with ethylene sebacate are also given. (Reproduced with permission from Ref. 4. Copyright 1968, Polym. Eng. Sci.)

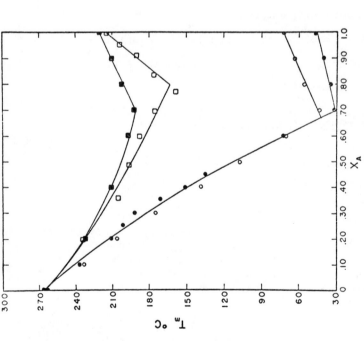

Figure 5. Compilation of melting temperature-composition relations for typical random copolymers and copolyamides. Key: ●, polyethylene terephthalate/adipate; ○, polyethylene terephthalate/sebacate; ■, polyhexamethylene adipamide/sabacamide; and □, polyhexamethylene adipamide/caproamide.

agreement with theoretical predictions. For the block copolymers the melting point remains constant for a large co-unit content and is independent of the chemical nature of the co-unit. Only when the co-unit contents become extremely large does the melting point decrease, consistent with the crystallization of the added species. The results that are shown in Figure 6 are quite typical of all types of block copolymers irrespective of their chemical constitution.

The large variation in melting temperature, that can be attained for a given composition, gives great versatility in the control of properties. For example, with block copolymers, a crystalline polymer can be modified without a significant reduction of its melting point, modulus, tensile strength and elongation. There is thus the possibility of maintaining desirable mechanical properties while enhancing other properties such as dyeability, water sorption, or elasticity by the proper selection of co-units. On the other hand, if a reduction of the melting temperature is required, perhaps for processing purposes, then the random introduction of the co-units would be desired.

Although homopolymers and block copolymers melt relatively sharply, random copolymers usually show very broad fusion ranges. This is a consequence of a greatly exaggerated impurity effect brought about by sequence length requirements. This phenomenon is illustrated in Figures 7 and 8 for a set of polybutadienes and polypropylenes, respectively. Although the repeating units are chemically identical for each of these polymers, they are properly treated as copolymers because of the structural irregularities incorporated within the chain. For polybutadiene, there is a varying content of the 1-4 trans crystallizing unit. As is shown in Figure 7, as the content of the crystallizing component decreases, the melting temperature is lowered and the fusion process becomes broader and becomes very difficult to detect as is evidenced by curve C. However, it is important to establish the existence of even small amounts of crystallinity because of its influence on mechanical and physical properties.

The melting of the stereoirregular polypropylene samples described in Figure 8 follow a similar pattern. The melting temperature and level of crystallinity decreases as the crystallizing isotactic content decreases. Concomitantly the fusion process becomes very much broader. It is not always a simple matter to be able to recognize the crystallinity and the fusion of such random type copolymers.

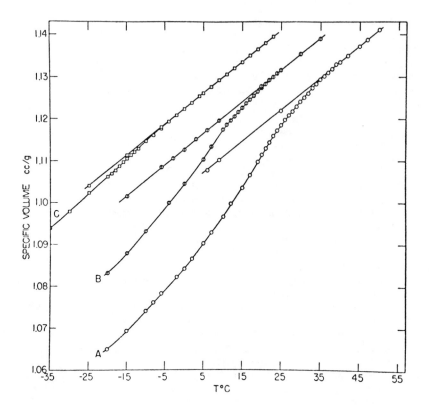

Figure 7. Fusion of random copolymers. Plot of specific volume against temperature for polybutadienes of different mole fraction, X_A, of crystallizing 1,4 trans units. Curve A, X_A = 0.81; curve B, X_A = 0.73; and curve C, X_A = 0.64. Curves B and C are arbitrarily displaced along the ordinate. (Reproduced with permission from Ref. 5. Copyright 1956, J. Polym. Sci.)

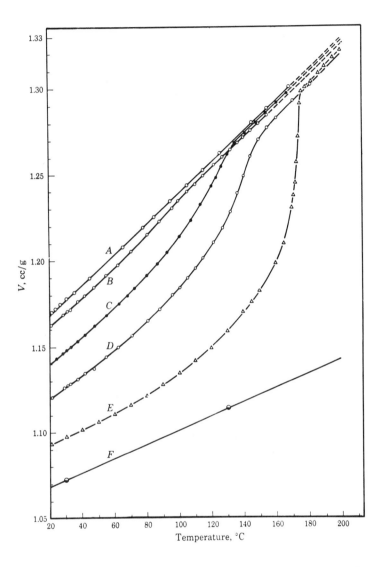

Figure 8. Fusion of random copolymers. The melting of polypropylenes of different stereoregularities. Plot of specific
volume as a function of temperature. Curve A, ether extract,
quenched; curve B, pentane extract, annealed; curve C, hexane
fraction annealed; curve D, trimethyl pentane fraction
annealed; curve E, experimental whole polymer annealed; and
curve F, calculated for pure crystalline polymer. (Reproduced
with permission from Ref. 6. Copyright 1960, J. Polym. Sci.)

Another important type of chain irregularity in this context is branching, since the branch points are structurally different from the other chain repeating units. An example of the effect of branching on the crystallization behavior is shown in Figure 9 for two polyethylene polymers. Curve A is for the linear polymer; curve B for the branched. The melting temperature of the branched polymer is significantly less than its linear counterpart. In addition, as would be expected, its melting range is very much broader. For the linear polymer most of the melting takes place over only a 3-4° interval. Thus we have a striking example of two very chemically similar polymers which display markedly different crystallization behavior.

By utilizing phase equilibrium theory it has also been possible to predict the melting temperature-molecular weight relations for fractions and for polydisperse systems which possess a most probable molecular weight distribution.

We summarize this section by concluding that there is a substantial body of evidence which demonstrates that formal phase equilibrium thermodynamics can be successfully applied to the fusion of polymers and has many far reaching consequences.

CRYSTALLIZATION KINETICS

A formal understanding of the processes involved in the crystallization kinetics from the pure melt has also been substantially developed. With only minor modification polymer crystallization has been shown to follow the general mathematical theory that was developed many years ago by Avrami for the crystallization of metals and other low molecular weight systems.

A typical set of crystallization kinetic isotherms for a pure polymer crystallizing from the melt is given in Figure 10. This example is for a molecular weight fraction of linear polyethylene. In this figure the extent of the transformation, or degree of crystallinity, is plotted against the log of time at different crystallization temperatures. Some very important features of the crystallization process are illustrated by these curves. The isotherms have a very characteristic sigmoidal shape, typical of all homopolymers. There is an initial induction time, which is more apparent than real, followed by a period of accelerated crystallization. A retardation of the crystallization process, sometimes called secondary crystallization, then occurs and a pseudo-equilibrium level of crystallization is reached. After sufficient time the same limiting

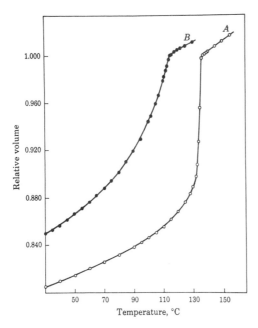

Figure 9. The effect of branching on the melting process. Plot of relative volume against temperature for linear polyethylene (curve A) and branched polyethylene (curve B). (Reproduced from Ref. 7. Copyright 1953, American Chemical Society.)

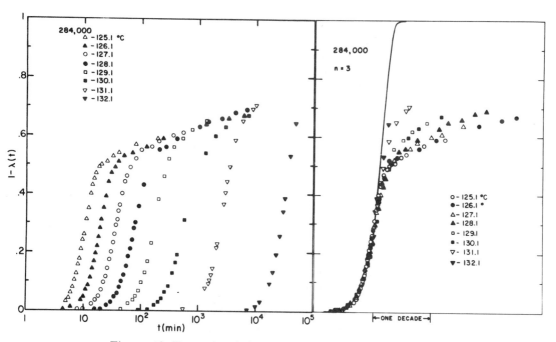

Figure 10. Example of the crystallization kinetics from the pure melt. Plot of degree of crystallinity against log time for a molecular weight fraction of linear polyethylene, M = 2.84 × 10⁵, at indicated temperatures. (Reproduced from Ref. 8. Copyright 1972, American Chemical Society.)

value is attained at each crystallization temperature. The time rate of change of crystallinity is extremely small in this region. Complete crystallinity is rarely, if never, attained and the level attained is molecular weight dependent.

A very strong negative temperature coefficient is also apparent from these plots. As the temperature is decreased the crystallization rate becomes much more rapid. This behavior is quite the opposite of what is found for the usual chemical reaction. The negative temperature coefficient is rather severe; in the example given the crystallization rate changes by five orders of magnitude over only a seven degree temperature interval. This behavior is clearly indicative of a nucleation controlled crystallization process. It illustrates an extremely important principle which underlies many aspects of polymer crystallization.

The isotherm shapes at each temperature appear to be very similar. They are in fact identical and can be superimposed one upon the other merely by shifting them along the horizontal axis. Thus one master isotherm results as is illustrated in the right-hand portion of Figure 10. This procedure shows that there is a single reduced time variable, dependent on temperature, which describes the crystallization process. The isotherm shapes and temperature coefficients illustrated in Figure 10 are typical of homopolymer crystallization.

Another popular, as well as useful, way by which to analyze the crystallization process is to study the spherulitic growth rate by direct light microscope examination. Several general features have emerged from the large number of such studies that have been made on many polymers. Spherulites grow at a linear rate and also display a nucleation type temperature coefficient. These observations are thus completely compatible with those for the overall crystallization process which are shown, for example, in Figure 10.

GENERAL ASPECTS OF STRUCTURE

Since the basic framework of the subject, comprised of the thermodynamics of fusion and crystallization kinetics, is well established we might quite logically and properly ask why there are still problems that remain to be resolved. One way to address this question is to consider the crystallization of a normal hydrocarbon.

For the normal hydrocarbons, even those which contain as much as a hundred carbon atoms per chain, it is well known that crystallization will take place

very rapidly by lowering the temperature only infini-
tesimally below the equilibrium melting temperature.
On the other hand, in order to crystallize the polymeric analogue, linear polyethylene, one has to reduce
the temperature well below the melting temperature
even for low molecular weight fractions. In the
former case molecular crystals are formed, since each
molecule is exactly the same length. In the latter
case they cannot develop even for the best fractionated samples, since there is a distribution of chain
lengths. Thus, the crystallization of long chain
molecules will occur at finite or reasonable rates
only at large undercoolings, i.e. from 20-40° below
the melting temperature. As a consequence, with
polymers one forms a polycrystalline system, which
is in fact only partially or semi-crystalline. The
crystallite and associated supermolecular structures
are complex. It is these morphological features
which determine the actual properties. The fact that
polymers can only crystallize at a finite rate under
conditions well removed from equilibrium presents
the basic problem. Therefore, in order to describe
and understand properties we are concerned with a
very morphologically complex non-equilibrium system.
These considerations bring us to the more modern
aspects of the problems involving the crystalline
state in polymers.

To put the problems with which we are faced into
some sort of perspective, we have prepared the chart
of Figure 11 which illustrates the interrelations
among the various subjects that are involved. Essentially, all properties are controlled by the molecular morphology. The molecular morphology is in turn
determined by the crystallization mechanisms. One
deduces such mechanisms from detailed studies of
crystallization kinetics. The equilibrium requirements are necessary in order to properly analyze the
kinetics. There is obviously a very strong inter-
relation among these different aspects of the problem.
Very little, if any, of the total problem concerned
with the crystallization behavior, or properties in
the crystalline state, can be examined in isolation.
Keeping these facets of the totality of the problem
in mind, we will focus our attention primarily on the
relationship between the molecular morphology and the
properties of homopolymers crystallized from the pure
melt. The principles that will be established, can
and have been extended to include polymer-diluent
mixtures and various type copolymers. Crystallization under an applied stress, or oriented crystallization, presents another, distinct area which will
not be discussed here.

Before proceeding, we should explain in more

detail what is meant in the present context by structure and morphology. The different levels of structure or morphology that are encountered in the study
of semi-crystalline polymers are illustrated by the
chart in Figure 12. The unit cell structures
are essentially the same as in the conventional
crystallography of low molecular weight substances.
The crystallite structure, resulting from the polycrystalline nature of the system, involves a description of both the internal structure of the crystallite, its associated interfacial region or zone, and
the interconnections, if they exist, between crystallites. The supermolecular structure also needs to be
considered. This is a concern for the organization
of crystallites into larger structures.

The determination of the unit cell structure is
treated by x-ray crystallographers in a classical
manner. It has not presented any real interpretive
problems. In contrast, the matter of the crystallite
structure has been a very controversial, and, unfortunately, divisive one for the last 20-25 years. A
rational analysis and resolution of this problem is,
however, finally at hand. Systematic work on the
supermolecular structure is just starting to evolve,
particularly the specification of the different types
of superstructures than can be developed under different conditions and their influence on properties.

It is important that we take note of the fact that
the molecular morphology differs in a very important
and significant way from what we might term the gross
morphology. Both of these concepts, however, are
clearly important. The gross morphology is what is
observed by direct microscopic examination; it describes the form or the shape of the structures of interest. The molecular morphology is a description of the
arrangement and disposition of the chain units which are
consistent with the gross morphology. Obviously, the
molecular morphology cannot be directly observed.

CRYSTALLITE
STRUCTURE

We direct our attention now to the problems of
crystallite structure. It is well established, and
accepted, that a lamellar-like crystal habit is the
characteristic, gross, morphological form developed
by homopolymers during crystallization from the pure
melt. Such lamellae were first observed for crystallites formed from dilute solution. The characteristic, typical thin lamellar habit is shown in Figure
13 for linear polyethylene. Such structures have
now been observed for all homopolymers studied and
they possess some very characteristic features. The
lamellar thickness, for dilute solution formed crystals,

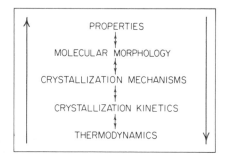

Figure 11. Perspective of crystalline state. Representation of problem areas in the study of crystalline polymers. (Reproduced with permission from Ref. 9. Copyright 1979, Faraday Discuss. Chem. Soc.)

UNIT CELL
CRYSTALLITE
SUPER MOLECULAR

Figure 12. Elements of structure describing semicrystalline polymers.

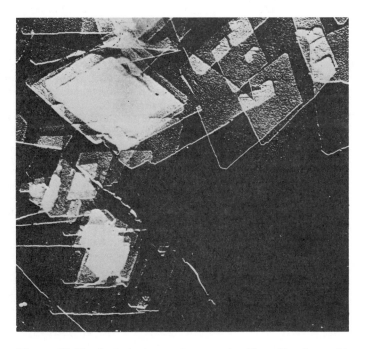

Figure 13. Typical electron micrograph of lamellae formed by homopolymers crystallized from dilute solution. Example illustrated for linear polyethylene.

is the order of 100–200 Angstroms, depending on the crystallizing solvent and temperature. The chain axes are preferentially oriented perpendicular to the basal planes of the lamellae. Since such crystal habits are found for very high molecular weight polymers and the thickness of the crystallites in the chain direction is only of the order of 100–200 Angstroms, it is obvious that a single chain must traverse the crystallite from which it originates many times. Consequently, the nature of the interfacial structure is quite important; it is not obvious, and cannot be deduced from the microscopic studies. It must be emphasized that despite the esthetic pleasantries of the crystallites shown in Figure 13, the interfacial structure is not at all apparent.

Although in this discussion we will not dwell in any detail on the properties of solution crystals, it is important to recognize that these kind of electron microscope observations do not lend themselves to a description of the interfacial structure on a molecular level. The gross morphological form and the orientation features are, however, well established. The interfacial structure, on the other hand, is consistent with several extremes as is schematically indicated in Figure 14. In one extreme, termed the regularly folded-adjacent reentry structure, the molecular chains appear to be accordian-like, making precise hairpin turns to yield the optimum level of possible crystallinity. Equally consistent with the gross morphology is the other model that is illustrated. Here, there is a distinct, disordered, amorphous overlayer. This model has been popularly termed the "switchboard" model. The reason for introducing these ideas is that a lamellar type crystallite is also the universal mode of homopolymer crystallization in the bulk.

The first lamellae that were observed in bulk crystallized systems, were obtained by surface replica electron microscopy. Unfortunately, the thicknesses of these lamellae were in the range of 100–200 Angstroms. A typical example of such lamellae crystallites in a bulk crystallized polymer is shown in Figure 15. These dimensions were originally thought to be typical of and unique to the lamellae formed during bulk crystallization. We know now that the lamellae thicknesses, depending upon molecular weight and crystallization temperatures, can range up to 1,000 Angstroms or more, even when crystallized at atmospheric pressure. There is, however, an apparent regularity to the gross morphological structures that are illustrated in Figure 15. Since the crystallite thickness in this early work is about the same

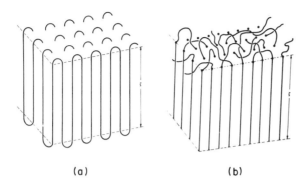

(a)   (b)

Figure 14. Schematic diagrams of possible chain structures within lamellar crystallite. Key: a, regularly folded array and b, nonregularly folded chains; loop lengths are variable. (Reproduced from Ref. 10. Copyright 1962, American Chemical Society.)

Figure 15. Lamellar structures of bulk crystallized polymers. Electron micrograph of surface films of linear polyethylene. (Reproduced with permission from Ref. 11. Copyright 1959, J. Polym. Sci.)

as that of solution crystals, a connection and iden-
tification between the two situations was immediately
made. It was proposed, in a very vigorous manner,
that the lamellar crystallites, observed in bulk
crystallized polymers, were comprised of regularly
folded chains which formed a smooth interface. Moreover, it was also thought that there were no molecular
connections between crystallites. To put matters in
another way, based principally on the gross morphological observations, the existence of non-crystalline
regions was disavowed.

It is, however, widely recognized, and we shall
describe many examples subsequently, that major
deviations exist in thermodynamic, as well as other
properties, from that of the perfect crystal. In
addition to the thermodynamic ones, these properties
include halos in the wide angle x-ray scattering patterns, the nature of the infrared and Raman spectra,
and the proton and carbon-13 nmr spectra. It was
thought that these deviations could be accounted for
by contributions from the smooth interface, since
small crystals are involved, as well as from a major
contribution from defects believed to exist within
the interior of the crystallite. A crystalline polymer
was viewed as consisting of disordered material, or
defects, imbedded within a crystalline matrix. Chain
units in non-ordered conformations, which would connect crystallites, did not exist in this view.

The implication of these ideas, or more pos-
itively stated, the establishment of the structure of
the crystallite on a molecular level is a very crucial
matter. It goes to the heart of the relationships
between structure, morphology and properties. Properties must eventually depend on the details of the
chain structure within the semi-crystalline polymer.
As we now begin our efforts to establish the crystallite structure certain basic principles need to be
recognized. It is important that these principles
be well understood. Otherwise they can be misused
and will inadvertently contribute to the confusion
and divisiveness of the problem.

Lamellae-type crystallites are now well recognized and universally accepted as the characteristic
mode for homopolymer bulk crystallization. Surprisingly, copolymers up to a relatively high co-unit
content also form lamellar crystallites. There are
some very important restraints to these observations
for homopolymers which must be made. The visual
observation of lamellae, upon which this generalization is based, or even the occasional viewing of
defined sectors within lamellae, is not a license to
describe the interfacial structure, the presence or
absence and nature of connecting regions, or even

the type and concentration of internal defects. The
apparent geometric regularity, as perceived by the
electron microscope, is obviously a gross morphological observation. It cannot by itself be taken as
evidence for the detailed structure. The resolving
power of the electron microscope, as applied to these
kind of problems, is not adequate to resolve these
details of chain structure. This fact, unfortunately,
has not always been recognized. Constant repetition
of the same argument does not validate it. The
observation of lamellae per se gives us no detailed
information about the disposition and arrangement of
the chains.

Another important point that must be considered
results from the fact that the crystallization kinetics of polymers follow nucleation kinetics. Hence
there is nucleation control of the crystallization
process. This is an extremely important concept; it
is a very crucial factor in the control of polymer
properties. As we indicated previously, the basis
for the deduction of nucleation control results from
the temperature coefficient studies of crystallization kinetics. This conclusion is reached on very
general grounds which are independent of the molecular
form or shapes of the nucleus. These observations
and conclusions do not therefore give one license to
select a specific molecular structure or form for the
nucleus. The kinetic theory of chain folding is
based on such a premise. The specifics of the nuclei
structure need to be established by methods other
than crystallization kinetics. Unfortunately, despite implications to the contrary, this goal has not
as yet been accomplished. Thus, no unique molecular
information can be obtained from the important deduction of a nucleation controlled crystallization
process.

We must also recognize that there is no scientific principle which requires the mature crystallites
which develop, and the nuclei from which they form
to have the same shape and molecular structure.
Major changes in the interfacial structure and in
shape can occur as a nucleus evolves into a mature
crystallite. The assumption that both species are
the same imposes an unnecessary, stringent restriction
on the problem and is usually incorrect.

The melting point-crystallization temperature
relations, or the related melting point-crystallite
size relations, that have widespread application in
polymer studies are fundamentally of a very general
nature. They are based on the theory of melting of
any type crystals of finite size. These relations
do not really give any unique information about
polymers, and specifically no information with regard
to the interfacial structure.

Despite these cautionary remarks, the general principles which we have described are very important in the analysis of polymer crystallization. However, proper recognition and understanding of their restrictions and limitations is equally important. Despite the widespread application of these general concepts, which has always involved a pre-assumption, there is no substantive experimental data to support the concept of regular folded interface or for the absence of chains units in disordered conformation. Any appeal that this model might still have must be esthetic in nature, since it rests on rather tenuous theoretical and experimental foundations.

With the establishment of these principles, and recognizing the precautions that must be taken, we can now examine the structural problem. Two different major approaches have been made to resolve this problem. One of these involves the study and analysis of how properties depend on molecular weight, using molecular weight fractions. Studies of bulk crystallized linear polyethylene, as well as poly(ethylene oxide), have shown that many properties are very dependent on the molecular weight. This phenomenon now appears to be quite general. Analysis of these results allows for the development of a molecular morphologic description of the elementary crystallite. We proceed now to examine these properties.

CRYSTALLITE STRUCTURE AND PROPERTIES

We consider first a very simple property, the density. Experimental results are presented in Figure 16, scaled as the degree of crystallinity, for molecular weight fractions of linear polyethylene crystallized to the fullest extent possible, at the relatively high temperature of 130°C, and never cooled. Hence, in these experiments the density measurements are made at the crystallization temperature. This is a very simple, uncomplicated experiment which is, however, quite rewarding. The results have far-reaching implications. The degree of crystallinity, or density, remains fairly constant up to molecular weights of about 10^5. As the molecular weight is increased further a monotonic decrease takes place in the density, with a leveling off occurring at molecular weights greater than 10^6. Densities at the high molecular weights correspond to degrees of crystallinity which are in the range of 0.25-0.30. Similar detailed studies show that poly(ethylene oxide) fractions behave in a very similar manner with molecular weight. The degree of crystallinity for this polymer approaches an asymptotic value of the order

0.20-0.25. Limited data with other polymers indicate similar trends. There is, therefore, a very definite dependence of the degree of crystallinity on molecular weight and the limiting degree of crystallinity that can be obtained in homopolymers.

The densities obtained after cooling the samples to room temperature, subsequent to crystallization, also are of interest. The results of such a study for linear polyethylene are shown in Figure 17 for two different modes of crystallization. One case is for isothermal crystallization and the other is for very rapid crystallization. The density is now seen to systematically depend both on the molecular weight and the crystallization conditions. For example, the densities, now measured at room temperature, range from 0.99, a value very close to that of the unit cell, to 0.94 following crystallization at 130°C and subsequent cooling. For linear polyethylene a density as low as 0.92 can be observed after the rapid crystallization of a high molecular weight fraction. After high temperature, isothermal crystallization and cooling the densities of the lower molecular weights approach those expected for the unit cell. The monotonic decrease in density now starts at a slightly lower molecular weight as compared to isothermal measurements. A constant value is reached in the very high molecular weight range. More rapid, non-isothermal crystallization results in much lower densities at comparable molecular weights. The molecular weight dependence of the density is no longer as severe. The main changes now occur at molecular weights which are less than about 10^5. For molecular weights greater than 10^5 only a small decrease in density is observed with increasing chain length.

A major conclusion from the density studies is that deviation from the unit cell value clearly cannot be attributed to chain ends within the crystal, or to effect of so-called "cilia". The utilization of cilia to explain many properties is periodically regenerated, although they are not supported by these very elementary facts. The largest deviations are observed as the concentration of chain ends is decreased with increasing molecular weight. Any significant role of chain ends, beyond very low molecular weights, is not at all obvious.

Although these density studies are important in themselves they also portend the results for many other properties. As an example, results of enthalpy of fusion measurements are summarized in Figure 18. For the ideal polyethylene system an enthalpy of fusion of 69 ± 1 calories per gram would be expected. In the figure, the observed values

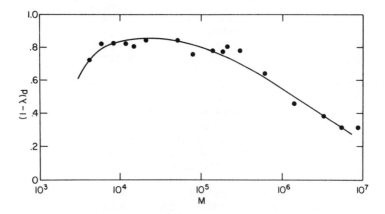

Figure 16. Plot of degree of crystallinity (from density) $(1 - \lambda)_d$, as a function of molecular weight for fractions of linear polyethylene crystallized at 130 °C and never cooled. Densities measured at the isothermal crystallization temperature. (Adapted from Ref. 8.)

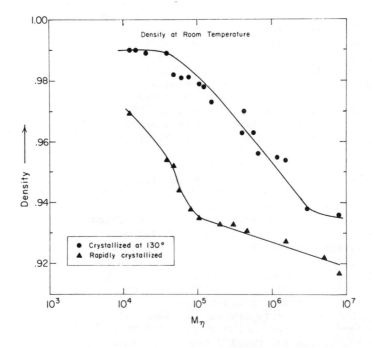

Figure 17. Plot of density, measured at room temperature, as a function of molecular weight for linear polyethylene fractions crystallized under the indicated conditions. (Reproduced with permission from Ref. 12. Copyright 1971, J. Phys. Chem.)

range from 68 to 37 calories per gram for the high
temperature crystallization and are reduced to the
range of 20 to 25 calories per gram after rapid crystallization. There is then a very wide range in
enthalpy values. The systematic dependence on molecular weight and crystallization conditions is very
similar to that observed for the density. Other
properties, such as the intensity of certain infrared
and Raman absorption bands, broad line proton nmr,
dynamic mechanical properties, wide angle x-ray diffraction and thermal expansion coefficients, also
show a very similar dependence on molecular weight
and crystallization conditions.

Certain important general features have emerged
from these types of properties studies. The deviations in values from that expected for the unit cell
(the perfect crystal) are systematic with molecular
weight and the mode of crystallization. These deviations are far from trivial. They are, as a matter
of fact, quite significant. It becomes clear that
one must examine many properties in order to develop
a complete, or meaningful, picture of the crystalline
state. A single piece of data, as for example an
isolated density value, can be interpreted in vir-
tually any arbitrary manner one desires. We must
recognize that there is no unique value to a given
property, because for the same polymer it will depend
on molecular weight. Focusing attention on an isolated piece of data can easily be a very treacherous
experience. Examining the complete set of data
imposes rather extensive, rigorous demands that must
be satisfied before any structural analysis can be
given to the crystalline state.

We now consider the crucial question as to what
the factors are which cause the deviations in properties from that of the unit cell. In a very elementary
way, one can state that the causes are to be found in
structures within the crystalline lattice, or by
structural features exterior to the crystal itself.
It is quite clear that one cannot put the burden on
the chain ends.

If the deviations in properties are caused in
any meaningful way by imperfections within the crystal
lattice, then their magnitude is such that changes in
the lattice parameters, i.e. the unit cell dimensions,
must occur. However, as is illustrated in Figure
19, it is found for linear polyethylene that as
the macroscopic density is varied over its extensive
range of 0.92 to 0.99, the actual lattice parameters
remain constant. We therefore must conclude that the
observed deviations of properties cannot be attributed in any significant way to a concentration of
imperfections within the lattice. Unfounded assertions

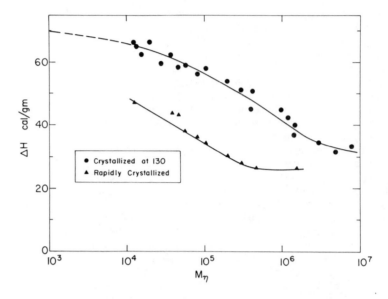

Figure 18. Plot of enthalpy of fusion as a function of molecular weight for linear polyethylene fractions crystallized under the indicated conditions. (Adapted from Ref. 12.)

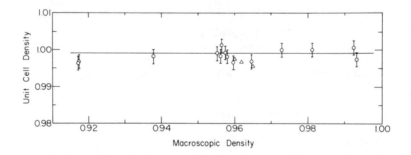

Figure 19. Linear polyethylene lattice parameters. Plot of unit cell density against macroscopic density for linear polyethylene fractions. (Reproduced with permission from Ref. 13. Copyright 1970, J. Polym. Sci., Polym. Phys. Ed.)

to the contrary have been made. The origin of the deviations, and the basis for the properties observed, must be sought in structures which are located outside of the crystalline region; that is, in structures which are external to the crystallite itself.

The investigation of this possibility resolves itself into a consideration whether the degree of crystallinity is a valid quantitative concept. Contributions to the properties from both the crystalline and the liquid-like regions needs to be taken into account. In addition one also must be concerned with possible contributions from the interfacial regions. This distinct possibility is suggested by the relatively high interfacial free energy that is associated with the basal planes of the lamellae-like crystallites formed from the pure melt. These interfacial free energies have been determined in a classical, straight-forward manner from independent studies of the relationship between the melting temperature and the crystallite thickness. An interfacial free energy of about 8,000 calories per mole of sequence is associated with the mature crystallites formed by high molecular weight linear polyethylene chains. This interfacial free energy is at least one, and possibly two, orders of magnitude greater than the interfacial free energy that is usually associated with crystals of low molecular weight substances.

In order to test the legitimacy of the degree of crystallinity concept it is necessary to use many physical methods. The reason is that different properties are sensitive to the different structural regions. The wide range of properties that can be generated by studying molecular weight fractions make such samples ideal to test the validity of the degree of crystallinity concept. This concept, of course, carries along with it the fact that the liquid-like regions co-exist with the ordered ones.

It has now been documented in great detail that the degree of crystallinity of linear polyethylene is indeed a very quantitative concept when the contributions from the interfacial regions are taken into account. The properties that have been studied which substantiate this statement include density, infrared absorption, enthalpy of fusion, wide-angle x-ray scattering, small-angle x-ray scattering, Raman absorption and broad line proton nmr among others. This quantitative conclusion about the degree of crystallinity is quite general and not limited to polyethylene. The classical work with natural rubber, where much lower levels of crystallinity are obtained, substantiates this conclusion. It has also been established for other polyolefins, for polyamides, polyesters and poly(tetrafluoroethylene) to cite but a few other examples. There is, therefore, a substantial body of experimental evidence which demonstrates that the degree of crystallinity is a quantitative concept applicable to all polymers. The earlier idea that amorphous, liquid-like regions do not exist is not correct. The deviations in properties from that of the unit cell must be primarily attributed to chain structures which are external to the crystalline regions. In a somewhat ironical note modern electron micrographs, based on thin sectioning and preferential staining techniques, also require the existence of liquid-like or amorphous regions. The well-defined structures that are observed in such micrographs result from the differential absorption of the stain between the crystalline and non-crystalline regions.

The results of the study of the properties of linear polyethylene, and other polymers as well, can be summarized by the fact that for molecular weights equal to greater than about 1×10^5 there is a continual decrease in the level of crystallinity until very high molecular weights are reached. This change is reflected in many of the properties that have been studied. At the same time, however, the crystallite thickness, the interfacial free energy associated with the basal plane, and the observed melting temperatures remain constant. These conclusions are based on very fundamental, direct experimental observations. These, coupled with the fact that the initial kinetics of the transformation are independent of molecular weight [see Figure 10] lead to the inescapable fact that there must be a significant, non-crystalline portion of chain units that are associated with semi-crystalline polymers.

From these studies the structure of an elementary crystallite can be developed which explains properties and their dependence on molecular weight and crystallization temperature. A schematic of such a rudimentary crystallite is illustrated in Figure 20. This schematic is meant to depict the structural highlights of the crystallite and is not concerned with the fine detail. Past experience teaches us that it would be prudent to first establish the existence of a forest before trying to identify the trees.

Three major regions, each having a different basic chain conformation is indicated in this schematic. These are the crystalline region, the interfacial region and the interzonal, amorphous, or liquid-like region. The crystalline region represents the three dimensional ordered structure with the typical lamellae-like habit. The imperfection levels within the crystallites formed by polymers are no

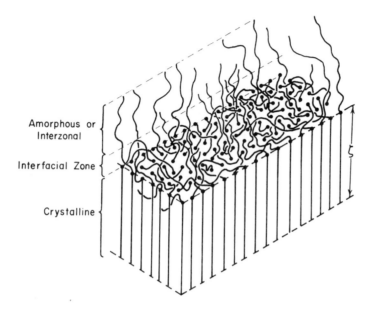

Figure 20. Schematic representation of a rudimentary crystal-lite. (Reproduced with permission from Ref. 12. Copyright 1971, J. Phys. Chem.)

different than those found in crystals formed by similar low molecular weight compounds. The crystallite, or core thickness as it is commonly called, is controlled by the nucleation requirements. However, the nucleation control does not mean regular folded chains. Neither does it require crystallite thicknesses which are identical with those of the critical nuclei. As we discussed earlier independent structural information about the nucleus is not available, nor can it be obtained solely from kinetic data.

The interfacial region is diffuse; it is not the sharp, clearly defined boundary that one usually associates with the interface of crystals of low molecular weight substances. This is an important distinction unique to chain molecules. This boundary is characterized by a very high interfacial free energy. There also is direct experimental evidence for the existence of the interfacial regions which can be quantitatively measured. For example the broad line proton nmr spectra cannot usually be decomposed solely into crystalline and amorphous regions. A third component, clearly involving the interface, is required to satisfy the experimental data.

An extensive set of experimental data utilizing this technique to study molecular weight fractions of linear polyethylene is given in Table II. In these results a wide range in the levels of crystallinity is attained by molecular weight control. There is good agreement in the degrees of crystallinity obtained from either broad line proton nmr or density for corresponding samples. As the molecular weight increases the level of crystallinity of course decreases; concomitantly, however, there is an increase in interfacial content from about 16 to 18 percent in these samples. The proportion of interfacial content increases with the higher molecular weight-lower crystallinity samples.

Another method which allows the interfacial content to be estimated involves the analysis of the internal modes of the polyethylene Raman spectrum. Some typical results obtained by this method are given in Table III. Again there is a significant interfacial region for the higher molecular weight samples. The values scale in a very similar manner to those obtained from the broad line nmr. However, slightly lower values are obtained by the Raman method. The differences obtained by the two methods is of relatively minor concern at present. The important point is that both methods yield significant interfacial contents which scale in a very similar way with molecular weight and crystallization conditions. Table III also gives the important

Table II. Broad Line Proton NMR for Polyethylene[a]

$M_\eta \times 10^3$	$(1-\lambda)_d$ [b]	From NMR		
		w_c [c]	w_a [d]	w_i [e]
13.3	0.931	0.939	0.002	0.059
31.8	0.922	0.919	0.002	0.079
90.0	0.868	0.870	0.035	0.095
150.0	0.826	0.822	0.063	0.115
248.0	0.761	0.755	0.091	0.154
431.0	0.755	0.740	0.109	0.151
1,000.0	0.674	0.677	0.155	0.168
3,400.0	0.639	0.636	0.187	0.177

(a) From Ref. 14
(b) Degree of crystallinity calculated from density
(c) Mass fraction crystalline
(d) Mass fraction amorphous
(e) Mass fraction interface

For a detailed discussion as to how these parameters are obtained see Ref 15.

Table III.   Raman Analyses of Polyethylenes

Linear Polyethylene

$M_w \times 10^3$	Crystallization Conditions	α_a (c)	α_c (d)	α_b (e)
5.6	123°C, 4 weeks	11 ± 4	89 ± 2	0 ± 6
27.8	130°C, 4 weeks	7 ± 3	92 ± 1	1 ± 4
70.0	131°C, 8 weeks	21 ± 3	82 ± 2	(-3) ± 5
188.5	130°C, 4 weeks	25 ± 3	69 ± 2	6 ± 5
428.0	126°C, 4 weeks	34 ± 3	55 ± 2	10 ± 5
1,620.0	130°C, 4 weeks	33 ± 3	56 ± 2	11 ± 5
27.8	quenched, -129°C	38 ± 3	56 ± 2	6 ± 5
188.5	quenched, -129°C	49 ± 3	46 ± 2	5 ± 5
316.0	quenched, -129°C	53 ± 3	35 ± 3	12 ± 6
428.0	quenched, -129°C	50 ± 3	39 ± 3	11 ± 6
1,620.0	quenched, -78°C	49 ± C	41 ± 3	10 ± 6

Branched Polyethylene

1.48×10^5 (6.1)[b]	quenched -78°C	45 ± 3	41 ± 2	14 ± 5
1.08×10^5 (22.0)	quenched -78°C	61 ± 5	23 ± 4	16 ± 9
5.1×10^4 (33.7)	quenched -78°C	76 ± 5	5 ± 1	19 ± 6

(a) From Ref. 16
(b) Branches/1000 carbon atom
(c) Percent amorphous
(d) Percent crystalline
(e) Percent interface

For a detailed description of the method used see Refs. 16 and 17.

result that there is a major increase in the interfacial content with branched polymers and copolymers.

Returning to our discussion of the schematic
crystallite, Figure 20, we find that the interzonal, or amorphous regions, involve chain units
which connect the crystallites. In this region the
chain units are in non-ordered conformations and
their properties are very similar to the completely
molten or random chains. This requirement is a
natural consequence of the quantitative nature of the
degree of crystallinity concept. In addition semi-
crystalline polymers display well-defined glass
temperatures. Glass formation is a property of the
liquid state. Its occurrence lends further support
to the presence of random structures. The sequences
of chain units connecting crystallites are not complete molecules. The term, "tie-molecules", which
has often been applied is a very bad misnomer. It
implies that the connections are extended or straight
and comprise a complete molecule. These connections
represent only portions of molecules and are clearly
in random conformation.

The schematic illustrated in Figure 20 represents the first step in the elucidation of the
structure of the crystallites. It is consistent with
all the observed properties and also explains how the
proportions of the different regions change with
molecular weight and crystallization conditions.
This model is completely compatible with direct
electron microscope observations as well as related
techniques. Despite the fact that non-crystalline
structures are not seen in the gross morphological
observations, their introduction is mandatory from
the consideration of properties.

As was indicated previously there is a second
major approach that can be taken in analyzing the
crystallite structure. A relatively new, and what has
has proven to be a highly informative method, involves the analysis of the small-angle neutron
scattering pattern of a mixture of hydrogenated and
deuterated chains in the semi-crystalline state.
The small-angle neutron scattering results that have
been obtained so far for several polymers yield,
with only minor exceptions, the same conclusions.
Two kinds of information have been obtained from
these experiments. The radius of gyration of a
chain in the crystalline state is obtained in the
usual way, without recourse to any models, by
extrapolating the scattering intensity to zero
angle. Quite surprisingly, as is shown in
Table IV, the molecular chain has the same

Table IV. Small Angle Neutron Scattering

Polymer	Method of Crystallization	$\dfrac{R_g^{(a)}}{M_w^{1/2}}$ $\left[\text{Å}/(\text{g/mol})^{1/2}\right]$	
		Melt	Crystalline
Polyethylene	Rapidly quenched from melt	0.46	0.46
Polypropylene	Rapidly quenched	0.35	0.34
	Isothermally crystallized at 139°C	0.35	0.38
	Rapidly quenched from melt and subsequently annealed at 137°C	0.35	0.36
Polyethylene oxide	Slowly cooled	0.42	0.52
Isotactic polystyrene	Crystallized at 140°C (5 hours)	0.26-0.28[b]	0.24-0.27
	Crystallized at 140°C (5 hours) then 180°C (50 min)		0.26
	Crystallized at 200°C (1 hour)	0.22[b]	0.24-0.29

Data from Ref. 18.

(a) R_g is radius of gyration of the molecule.

(b) Dimensions in the melt were not available. The values quoted are for amorphous samples (with atactic polystyrene matrices) annealed in the same way as the crystalline material (with isotactic polystyrene matrices).

For further details of the method and analysis see Ref. 19, 20, and 21.

radius of gyration in the pure melt and in the semi-crystalline state. From these results it becomes quite apparent that the lamellar crystallites do not contain any significant concentration of regular folded chains. The molecules do not crystallize as individual entities into regular structures. If this process occurred then the radius of gyration would be very different from what is observed. The virtual identity of the radius of gyration in the two states indicates that there has not been much readjustment in chain conformation as the crystallizing growth front advances.

The other information is obtained from the scattering intensity at the intermediate angles. Analysis of this type of data involves the assumption of some type of model. It is shown that there cannot be very much adjacent reentry of the crystalline sequences, if there is any at all. Much of the current discussion in this particular area has been centered on the limiting small number of such sequences that can be tolerated by the neutron scattering data. This discussion, although of some importance, has unfortunately had the tendency to obscure the major results of the neutron scattering studies. The unequivocal conclusion has been reached that regular folding does not occur to any meaningful extent. Such structures contribute very little to the properties and behavior of crystalline polymers.

There are, therefore, two distinctly different ways of approaching the problem of crystallite structure. On the one hand, there is the study of the wide variety of properties; and on the other the analysis of the neutron scattering patterns. Both types of studies result in very similar conclusions, namely, that the lamellar-like crystallites do not contain regularly folded chains, or compact structures that even closely resemble them. A crystallite structure of the general type illustrated in Figure IV.F.5 is favored.

There are, however, a variety of problems with respect to crystallite structure that still remain to be resolved. Most important among these are the fine structures of the different regions represented in Figure 20. These include a detailed conformational analysis of the interfacial region and its deformability; a more quantitative analysis of internal imperfections; and an analysis of the minor alterations that might occur from the pure liquid structures because units are anchored in the crystalline regions.

Recent structural studies have shown that the extent of the lateral development of the crystallite depends on molecular weight and crystallization temperature. Significant curvature is also developed within the lamellae, depending on these variables. This work strongly suggests a definite connection between the lateral size of the lamellae, their curvature and the supermolecular structure that develops.

SUPERMOLECULAR
STRUCTURE

Another major interest with respect to the crystalline state is the nature of the supermolecular structure. We are concerned here with the arrangement of the individual lamellar crystallites into a larger scale of organization. This aspect of structure has been an extensively studied and a widely reported phenomenon. We are well aware that such higher orders of organization exist. These structures manifest themselves, to take a specific example, in the common observation of spherulites in semi-crystalline polymers. Despite the widespread observation of such structures it has only been recently that they have been studied in any systematic way. Of primary importance and interest are the conditions under which different kinds of supermolecular structures are formed and their influence on properties. A powerful technique that can be used in these studies is small-angle light scattering. The light scattering studies are usually complemented by direct light and electron microscopic observations.

The most useful patterns for describing the superstructure are the H_V, which are dependent on orientation fluctuations. In this mode the incident light is polarized in the vertical direction and the observed scattered light is polarized in the horizontal direction. Although we shall only discuss the results for polyethylene, they are also similar to what has been found in molecular weight fractions of poly(ethylene oxide). Thus, the results described for polyethylene can be taken to be quite general.

The polyethylenes display five distinctly different types of light scattering patterns, which are illustrated in Figure 21 and are designated by the letters of the alphabet. These patterns range from that of the classical cloverleaf (a), to one which is circularly symmetric (h). The light scattering patterns can be related by theory to different supermolecular structures which are listed in Table V. In this Table, patterns (a) to (c) represent spherulites of decreasing order, that is, the spherulitic structure is deteriorating. Pattern (a), the classical cloverleaf, with zero intensity in the center represents the ideal, best developed spherulite. Pattern (d), which has an azimuthally dependent light scattering pattern, represents lamellae

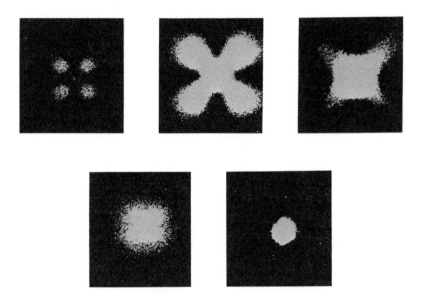

Figure 21. Types of light scattering patterns observed with polyethylenes. (Reproduced with permission from Ref. 9. Copyright 1979, Faraday Discuss. Chem. Soc.)

Table V. Light Scattering and Supermolecular Structure

Small Angle Light Scattering	Supermolecular Structure
a, b, c (cloverleaf)	a, b, c type spherulites
d (some azimuthal dependence)	d thin rods or rod-like aggregates
	g rods (breadth comparable to length); sheet-like
h (no angular dependence)	h randomly oriented lamellae

which are organized into thin rods or rod-like
aggregates. The circularly symmetric pattern, designated (h), does not represent a unique morphological
situation. It can represent rods, or sheets, where
the breadth is comparable to the width. We designate
this structure as a (g) type morphology. This same
light scattering pattern can also represent a random
collection of lamellae, which have no correlation
between them. We designate this structural situation
as an (h) type morphology. Hence the (h) type scattering pattern can represent either one of two kinds
of superstructures. These can only be discriminated
by the application of some complementary microscopic
method.

It can be stated in a general way that the supermolecular structures that are formed depend on the
molecular weight, the crystallization conditions,
such as the isothermal crystallization temperature
or the cooling rate, the molecular constitution and
the molecular weight polydispersity. The supermolecular structures are uniquely sensitive to polydispersity.

It is possible to establish a morphological map
which depicts how the supermolecular structure depends
on molecular weight and crystallization conditions.
We first consider only isothermal crystallization
conditions. The limits of the isothermal temperature
range can be easily established by kinetic experiments. The isothermal morphological map for linear
polyethylene, typical of other systems, is given in
Figure 22. In this Figure the dashed line delineates the boundary for isothermal crystallization.
The regions where the different superstructures are
formed are given the letter designation of Table
V. One of the immediate highlights of this map
is the fact that spherulitic structures are not always
observed. They are not the universal mode of homopolymer crystallization that has until recently been
our general understanding. In fact, as the map
clearly indicates, superstructures do not always
develop. An (h) type morphology is found, under
isothermal crystallization conditions, for high molecular weight fractions. When the map of Figure 22
is examined in more detail we find that the low molecular weight polymers form thin rod-like structures.
As the molecular weight is increased a (g) type morphology is observed at the higher crystallization
temperatures. Here the length and breadth of the
rod-like structures are comparable to one another so
that sheet-like structures are observed. If the isothermal crystallization temperature is lowered then
in this molecular weight range, spherulites will form.
The spherulitic structure deteriorates as the chain

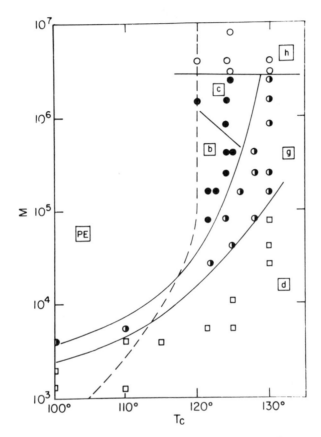

Figure 22. *Morphological map for isothermally crystallized molecular weight fractions of linear polyethylene. Supermolecular structures are designated as in Table V.*

length increases in this range. For molecular weights greater than about 2×10^6 no organized superstructures are observed at all, although the crystallinity level is the order of 0.50 to 0.60 for these samples.

A morphological map for the non-isothermal crystallization of linear polyethylene is given in Figure 22. In these experiments the temperature listed is not the crystallization temperature but that of the quenching bath to which the sample is rapidly transferred from the melt. Although subjective, this experiment is a reproducible one, which accomplishes the main purpose of varying the superstructure. The isothermal region, shown in Figure 23, has now been reduced to a narrow portion of the righthand side of the diagram. The non-isothermal portion does, however, merge in a continuous manner with the isothermal region. Although no new structural forms are observed, major differences can again be developed depending on the molecular weight and quenching temperature. The random-type (h) morphology can now be formed for molecular weights as low as 1×10^5 for cyrstallization after relatively rapid cooling. Hence this structure is not limited to the very high molecular weights. In contrast, well-developed (a) type spherulites can also be generated at low temperatures for very low molecular weight fractions.

By examining the map of Figure 23 we find that it is possible to prepare different supermolecular structures from the same molecular weight by choosing the appropriate crystallization conditions. In certain situations the different superstructures can be formed with the same molecular weight at the same level of crystallinity. There is, therefore, another well-defined independent variable which must be taken into account in discussing the properties and behavior of crystalline polymers.

Thin section, transmission electron microscopy, when properly carried out, can be very useful in studying the structure of crystalline polymers. Hence it is of interest to compare the morphological structures obtained from the small-angle light scattering studies with those from electron microscopy. A typical set of electron micrographs are given in Figures 24-26. In Figure 24 the sheet-like structures, or (g) type morphology, obtained from a fraction $M = 1.89 \times 10^5$, which was isothermally crystallized at $131.2^{\circ}C$ is given. For the same molecular weight sample, when quenched at $100^{\circ}C$, well developed spherulites are observed as is shown in Figure 25. Random lamellae, or (h) type structures, are illustrated in Figure 26 for $M = 6 \times 10^6$ crystallized at $130^{\circ}C$. Other structures can also be demonstrated by this electron microscopic

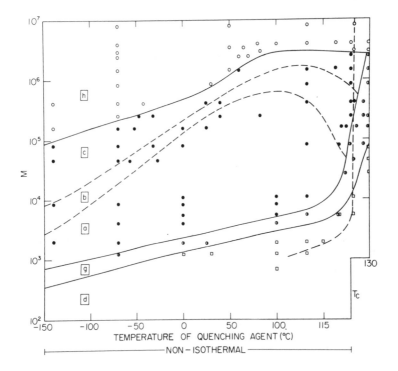

Figure 23. Morphological map for molecular weight fractions of linear polyethylene. Plot of molecular weight against either quenching or isothermal crystallization temperature. Supermolecular structures designated as in Table V. (Reproduced from Ref. 22. Copyright 1981, American Chemical Society.)

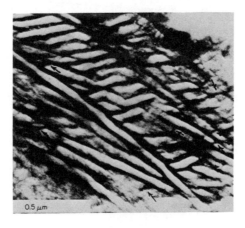

Figure 24. Transmission electron micrograph of linear polyethylene fraction $M_w = 1.89 \times 10^5$ crystallized at 131.2 °C. (Reproduced with permission from Ref. 23. Copyright 1981, J. Polym. Sci., Polym. Phys. Ed.)

Figure 25. Transmission electron micrograph of linear polyethylene fraction $M_w = 1.89 \times 10^5$ quenched at 100 °C. (Reproduced with permission from Ref. 23. Copyright 1981, J. Polym. Sci., Polym. Phys. Ed.)

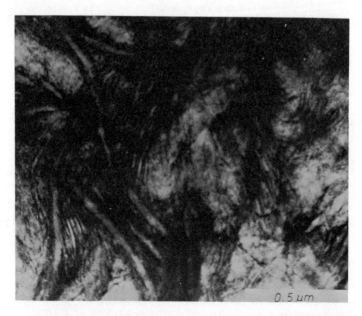

Figure 26. Transmission electron micrograph of linear polyethylene fraction $M_w = 6.0 \times 10^6$ crystallized at 130.0 °C. (Reproduced with permission from Ref. 24. Copyright 1980, J. Polym. Sci., Polym. Phys. Ed.)

technique. Comparison of the two methods shows that there is a one-to-one correspondence between the morphological map deduced from small angle light scattering and direct electron microscopic observation.

It is also of interest to examine the influence of the supermolecular structure on properties. We consider the relationship between the density, and thus the degree of crystallinity, and the supermolecular structure in Figure 27. Plotted here are the densities for different cyrstallization temperatures for a set of molecular weight fractions. The different supermolecular structures that are formed are also indicated. There are no morphological changes for $M = 10^4$ and the density changes smoothly with crystallization temperature. For molecular weights of 10^5 and 10^6 there are major changes in the supermolecular structure. However, these structural changes are not reflected in any changes in the density under comparable crystallization conditions. The enthalpy of fusion and the measured melting temperature show a similar insensitivity to the supermolecular structure. The superstructure thus has very little influence on the thermodynamic quantities. Studies of different types of absorption spectroscopy also show that there is very little, or no, effect of the supermolecular structure. The type and size of the supermolecular structures must obviously affect optical properties. Beyond this, no definite influence of the superstructure on properties has as yet been established.

We also want to examine the influence of the chain constitution on the supermolecular structure. The incorporation of branched (side) groups, copolymeric units or other type irregularities into the chain alter the major characteristics of the morphological map. As a general proposition, as far as the supermolecular structure is concerned, a branched polymer behaves as though it were of a higher molecular weight relative to the results obtained for linear polymers. Analysis of isothermally crystallized samples for these polymers is complicated by the limitation of the small amounts of crystallinity that develop during crystallization and its influence on the major amount that develops on cooling. To avoid such complexities, we limit ourselves here to the reproducible, but non-isothermally crystallized, systems.

A typical morphological map. obtained in the standard way, is given in Figure 28 for a set of molecular weight fractions of branched polyethylene. Each of these fractions contains about 1.5 mol percent branch groups. For a given branching content

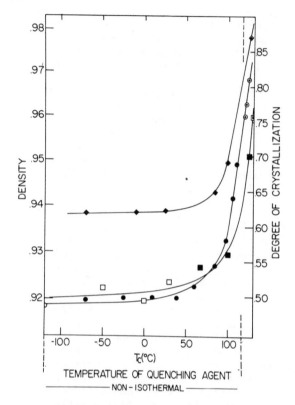

Figure 27. Plot of density, or degree of crystallinity, as a function of the isothermal crystallization temperature for three linear polyethylene fractions. Morphological forms are described for Figure 28. For $M_w = 2.78 \times 10^4$, spherulites ◆, rods (d) ◇; for $M_w = 1.61 \times 10^5$, random lamellae ◯; spherulites ●; rods (g) ⊙; for $M_w = 1.50 \times 10^6$, random lamellae ☐; spherulites ■; rods (g) ⊡. (Reproduced with permission from Ref. 9. Copyright 1979, Faraday Discuss. Chem. Soc.)

and molecular weight there is a very limited temperature range within which spherulites of differing degrees of order can form. Lower molecular weights are conducive to the formation of more highly ordered spherulites. When the superstructure is examined as a function of molecular weight a dome-shaped curve develops which forms the boundary for spherulite formation. For both higher and lower temperatures outside the dome boundary, the (h) type morphology of random lamellae is usually developed. Within the dome, spherulites are formed. This morphological conclusion is confirmed by thin section, transmission electron micrographs. For the branched polymers there is a very large molecular weight range over which lamellae can be formed.

Based on experiments with fractions of varying branching contents and molecular weight a schematic

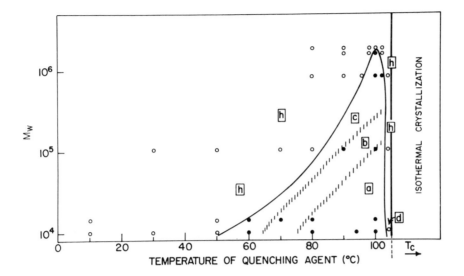

Figure 28. Morphological map for molecular weight fractions of branched polyethylenes, 1.5 mol% branched groups. Solid line delineates region of spherulite formation. Supermolecular structures are designated as in Table V. (Reproduced from Ref. 22. Copyright 1981, American Chemical Society.)

representation of the changes that take place in the boundary for spherulite formation is given in Figure 29. For a given molecular weight, as the branching concentration decreases, the temperature range over which spherulites can be formed becomes larger. The height of the dome, which encloses the spherulites, decreases with increasing branching content. Thus both increased molecular weight and branching content reduce the possibility of spherulite formation and favor the random arrangement of the lamellae.

The thermodynamic properties of the branched polymers, such as the density, enthalpy of fusion and melting temperature have been found to be independent of both the superstructure and the molecular weight. This last observation is distinctly different from what has been found for the linear homopolymers. Some typical density and enthalpy of fusion results are given in Figure 30 for three branched polyethylene fractions with the same branching content. These samples cover a very large molecular weight range and each contain about 1.5 mol percent of branches. These properties are governed primarily by the branching (or co-unit) concentration. The large changes in properties that are expected with molecular weight, based on the study of the linear analogues, are not in evidence for these type molecules.

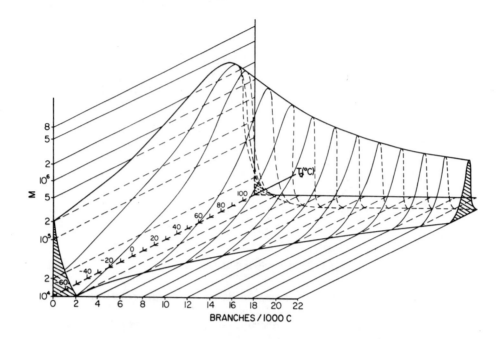

Figure 29. Three-dimensional schematic morphological map of the nonisothermal crystallization of the polyethylenes. The curved, dome-shaped regions define the volume within which spherulitic structures are formed; outside this volume there is' no defined supermolecular structure. (Reproduced from Ref. 22. Copyright 1981, American Chemical Society.)

CONCLUSIONS AND FUTURE DIRECTIONS

As our major objective, we have developed the basic thermodynamic, kinetic and structural principles which govern the crystallization behavior of polymers. These principles can be applied to develop an understanding of a variety of properties of semi-crystalline polymers. The phenomenological description of crystallization kinetics and the thermodynamic aspects of the subject have already reached a high level of comprehension and maturity. In contrast, an understanding of the structural features, at all the different hierarchies, has been in a state of rapid development. Because of the non-equilibrium character of the crystalline state in polymers, and the attendant morphological complexities, the influence of the structure is overriding. In addition to describing the different elements of structure in molecular terms, the role of chain constitution in determining the structure and influencing the properties and behavior needs to be established.

The overall problem can be looked at in the following way. There are a set of independent variables, which in the analysis of any given problem, need to be identified, isolated and quantified. These vari-

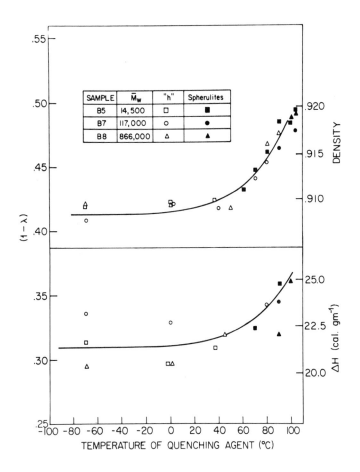

Figure 30. Plot of densities, enthalpies of fusion, and degree of crystallinity as a function of quenching temperatures for molecular weight fractions of branched polyethylenes containing 1.5 mol% branch groups. (Reproduced from Ref. 22. Copyright 1981, American Chemical Society.)

ables include the level of crystallinity, the crystallite thickness, the supermolecular structure, the interfacial content and structure, and the molecular weight. Different elements of chain structure, as well as the crystallization conditions, determine these variables, which in turn affect properties. The molecular weight, although determining some of these variables, is also important by itself. It determines the residual structure, or topology, of the non-crystalline region. The structure of this region plays an important role in determining mechanical properties. As we have seen, the level of crystallinity has a direct and important influence on many properties. The important role of the crystallite thickness and interfacial structure in influencing properties, is just beginning to be understood. Methods are now available to experimentally determine all of these variables. The resolution of many complex problems, including those of the ultimate properties, resides in this general kind of approach. The independent variables are varied over the widest extent possible, by control of molecular constitution and crystallization conditions, and the change in the property of interest observed. By following this strategy problems can be reduced to their important structural features and in this way resolved.

LITERATURE CITED

(1) F. A. Quinn, Jr. and L. Mandelkern, J. Am. Chem. Soc. 83 2857 (1961).

(2) L. Mandelkern, "Crystallization of Polymers" McGraw-Hill (1969).

(3) B. Wunderlich, "Macromolecular Physics" V.3, Academic Press (1980).

(4) J. F. Kenney, Polymer Engineering and Science 8 216 (1968).

(5) L. Mandelkern, M. Tryon, and F. A. Quinn, Jr., J. Polym. Sci. 19 77 (1956).

(6) S. Newman, J. Polym. Sci. 47 111 (1960).

(7) L. Mandelkern, M. Hellman, D. W. Brown, D. E. Roberts and F. A. Quinn, Jr., J. Am. Chem. Soc. 75 4093 (1953).

(8) E. Ergoz, J. G. Fatou and L. Mandelkern, Macromolecules 5 147 (1972).

(9) L. Mandelkern, Farad. Disc. of Chem. Soc. 68 310 (1979).

(10) P. J. Flory, J. Amer. Chem. Soc. 84 2857 (1962).

(11) R. Eppe, E. W. Fischer and H. A. Stuart, J. Polym. Sci. 34 721 (1959).

(12) L. Mandelkern, J. Phys. Chem. 75 3920 (1971).

(13) R. Kitamaru and L. Mandelkern, J. Polym. Sci. Part A-2 8 2079 (1970).

(14) R. Kitamaru, F. Horii, S. H. Hyon, J. Polym. Sci., Polym. Phys. Ed. 15 821 (1977).

(15) R. Kitamaru, Adv. in Polymer Sci. _26_ 139 (1978).

(16) M. Glotin and L. Mandelkern, Colloid and Polym. Sci. _260_ 182 (1982).

(17) G. R. Strobl and W. Hagedorn, J. Polym. Sci., Polym. Phys. Ed. _16_ 1181 (1978).

(18) G. D. Wignall, L. Mandelkern, C. Edwards and M. Glotin J. Polym. Sci., Polym. Phys. Ed. _20_ 245 (1982).

(19) J. Schelten, D. G. H. Ballard, G. D. Wignall, G. W. Longman and W. Schmatz, Polymer _17_ 751 (1976).

(20) D. Y. Yoon and P. J. Flory, Polymer _18_ 509 (1977).

(21) M. Stamm, E. W. Fischer and M. Dettenmaier, Disc. Farad. Soc. _68_ 263 (1979).

(22) L. Mandelkern, M. Glotin and R. A. Benson, Macromolecules _14_ 22 (1981).

(23) I. G. Voigt-Martin and L. Mandelkern, J. Polym. Sci., Polym. Phys. Ed. _19_ 1769 (1981).

(24) I. G. Voigt-Martin, E. W. Fischer and L. Mandelkern, J. Polym. Sci., Polym. Phys. Ed. _18_ 2347 (1980).

We suggest the following for supplementary reading and a more detailed description of the subject matter:

L. Mandelkern, "Crystallization of Polymers"; McGraw-Hill, 1964.

B. Wunderlich, "Macromolecular Physics"; Academic Press, 1980.

Faraday Discussions of the Chemical Society, "Organization of Macromolecules in the Condensed Phase; No. 68, 1979.

J. H. Magill in "Treatise on Materials Science and Technology"; J. M. Schultz, Ed.; Vol. 10, 1977, p. 3.

A. Keller Rept. Progr. Phys. 1968, _31_, 623.

L. Mandelkern, in "Progress in Polymer Science" A. D. Jenkin, Ed.; Pergamon Press, 1970; _2_, p. 165.

J. D. Hoffman, et al. J. Res. Natl. Bur. Stand. Sect. _A_ 1975, _79_, p. 671.

L. Mandelkern, "Morphology of Semi-crystalline Polymers" in "Characterization of Materials in Research: Ceramics and Polymers"; Syracuse University Press, 1975.

P. J. Flory, J. Chem. Phys. 1949, _17_, 223.

P. J. Flory Trans. Farad Soc. 1955, _51_, 848.

P. J. Flory J. Amer. Chem. Soc. 1962, _84_, 2857.

P. J. Flory and A. Vrij J. Amer. Chem. Soc. 1963, _85_, 3548.

P. J. Flory and D. Y. Yoon Nature 1978, _272_, 226.

J. Maxfield and L. Mandelkern Macromolecules 1977, _10_, 1141.

5

Molecular Spectroscopy

Jack L. Koenig

Most research concerning synthetic polymers is aimed
toward improving their useful material properties.
To do this systematically requires an understanding
of molecular-level influences on macroscopic behavior. Chemical structure, morphology, orientation,
and chain dynamics are known to have important
effects on bulk material properties, and these have
all been characterized by various forms of molecular
spectroscopy. Despite such efforts a complete understanding of the relationship between macroscopic and
microscopic properties does not yet exist.

C-13 NMR has become a powerful method of characterizing polymers in solution because of its simplicity
of interpretation, sensitivity to subtle molecular
changes, and relative ease of quantitation (1-3).
Much of this success is due to the sharpness of the
resonances, with linewidths typically on the order
of a few Hz or less. The rapid molecular motion in
solutions averages line-broadening interactions
which would otherwise dominate the spectra. This
averaging is much less effective in solid polymers
below their glass transition temperatures; the rates
and amplitudes of molecular rotation and translation
are substantially reduced. Linewidths for solid
polymers recorded with the same conditions as used
for solutions may be in excess of 20 kHz. These
linewidths, despite their magnitude, can provide
useful information. Broadline measurements of proton
dipolar linewidths have been used to determine self-
diffusion and heterogeneity in polymers, as well as

HIGH-RESOLUTION
C-13 NMR OF SOLID
POLYMERS

0851/84/0207$09.00/1
© 1984 American Chemical Society

to assess proposed models of chain motion (4,5).
However, much additional information is obscured by
the large linewidths resulting from incompletely
averaged interactions in the C-13 spectra. The
objective of the high-resolution techniques is to
selectively average these interactions and to thereby
obtain more detailed information concerning the
polymeric solid. The most common high-resolution
C-13 work involves the following procedures:
a) High-power decoupling; an essential part of the
experiment is the removal of dipolar coupling between
carbons and protons, which is generally the major
source of resonance broadening in polymers. High-
power radio frequency (rf) fields applied near the
proton Larmor frequency coherently average the
dipolar and scalar couplings between carbons and
protons (6). As far as the dipolar interaction is
concerned, this is the same result as is achieved by
stoichastic averaging via rapid molecular motion in
solutions. The rf power levels used for solid polymers are typically 1-2 mT (10-20 G), an order of
magnitude greater than those used for scalar decoupling in solutions. These rf field strengths are not
only greater than the C-13/H-1 couplings but also
greater than the H-1 dipolar linewidth, a condition
which is necessary for complete decoupling to occur
(7). In the high-resolution C-13 studies that we
will consider, high-power decoupling is always used;
the decoupler is switched on during the acquisition
period and then off for a delay which permits spin-
lattice relaxation and cooling of the probe, b)
Magic-angle spinning; even after the coupling to
protons is removed, the C-13 linewidths of solid
polymers are usually a few kilohertz in width. The
chemical shift anisotropy of the carbons is commonly
the largest contributor to the remaining linewidth.
Because of the anisotropic electronic structure surrounding a given carbon, the nuclear shielding will
be different for each orientation of the electron
cloud with respect to the static magnetic field. The
result is a broad resonance for each chemical type
of carbon in an amorphous or polycrystalline sample.
These resonances frequently overlap in the spectra of
solid polymers, making analysis difficult. However,
if the sample is rotated rapidly about an axis
oriented at 54.7° from the static magnetic field,
broadening from the chemical shift anisotropy is
removed (8). Magic-angle spinning, as this technique
is known, effectively averages the full chemical
shift tensor to its trace; the observed resonance
then appears at its isotropic value. Similar results
occur in solution by stochastic averaging through
rapid molecular motion. The reason that rotation

about a single axis is sufficient in solids is that the chemical shift interaction is dependent upon the time average of $3 \cos\theta - 1$, where θ is the angle between the rotation axis and the static field. When $\theta = 54.7°$, the magic-angle, the broadening term is eliminated leaving one additional term, the isotropic chemical shift. A few additional points are relevant here. First, if the rotational frequency is not greater than the chemical shift anisotropy, spinning sidebands will be observed. They are displaced from the isotropic chemical shift in the spectrum by multiples of the rotational frequency. Second, the dipolar interaction between spin-active nuclei also has the $3 \cos^2\theta - 1$ dependence (9,10). Magic-angle spinning should, therefore, eliminate dipolar broadening as well. It has been effective in doing so in selected cases (8). However, for most solid polymers elimination of C-13/H-1 dipolar coupling would require spinning frequencies on the order of 20 kHz, which is not routinely possible. In these cases rf decoupling of the protons is necessary. Finally, while the simplifications resulting from magic-angle spinning are often required in polymer analysis, there are instances in which nonspinning spectra are desirable. Static samples can give information concerning molecular motion, orientation, and carbon-carbon coupling which may be valuable. Isotopic enrichment at a particular carbon site may be used to lessen the problem of overlap, c) Cross-polarization; although the large linewidths are eliminated by simultaneous use of high-power decoupling and magic-angle spinning, high-resolution NMR of solid polymers still has a problem with low sensitivity. For conventional observation of the carbon spins in a Fourier transform experiment, the delay between successive acquisitions should be on the order of the C-13 spin-lattice relaxation time, T_1. For solid polymers such delays may be 1000 s or longer (11). These times can be shortened considerably by relying on the proton magnetization to relax to equilibrium and by then transferring this magnetization to the carbon spins. Cross polarization is the term used for this process. The proton T_1's are much shorter than the carbon ones because of the effectiveness of spin diffusion; those protons in a mobile, quickly relaxing region of the sample can cool those in more rigid areas through energy-conserving spin flips. The low natural abundance of the carbon spins prevents effective spin diffusion among themselves. Once the proton spins have relaxed, magnetization can be transferred to the carbons by a sequence of rf pulses whose power levels are adjusted to satisfy the Hartmann-Hahn match (12):

$$\gamma_c H_{1c} = \gamma_H H_{1H}$$

where γ represents the gyromagnetic ratio and H_1 designates the rf field power. In this condition both the carbon and proton magnetizations along the static field direction are being modulated at the same frequency. The result is a net transfer of polarization from the cold proton reservoir to the hot carbon spins. In addition to reducing the delay time between successive acquisitions, cross polarization offers a theoretical sensitivity enhancement per acquisition of a factor of 4 (13).

Some aspects of cross polarization are particularly important in its applications to solid polymers. The first of these is the quantitative accuracy of the experiment. Since magnetization is transferred from protons to carbons, reliable measurement of the number of carbons of all chemical types requires that the polarization efficiency be the same for each kind of carbon. Since there are variations in the transfer rates between protons and carbons (dependent on the separation distance) and variations in the proton reservoir throughout the sample, quantitative conditions are not always reached. Altering of experimental conditions may improve the situation, however, as will be discussed in later sections. Secondly, to obtain the maximum resolution and maximum sensitivity in the C-13 spectra of solid polymers, high-power decoupling, magic-angle spinning, and cross polarization are used simultaneously (14). In some materials the combination of magic-angle spinning and cross polarization is not helpful, since the static dipolar interactions required for polarization transfer are averaged by spinning (15). However, with polymers this is generally not a problem because the frequency of the proton-proton dipolar fluctuations is sufficiently high so that spinning frequencies of a few kHz have little effect. The techniques can then be used simultaneously without difficulty for polymer samples. Finally, there may be instances when cross polarization is not desirable. To separate spectroscopically the components of a structurally heterogeneous polymer system, differences in the C-13 T_1's may be used. One part of a system may have sufficient molecular motion for its C-13 spins to relax quickly and may then be isolated from more rigid components. This is of particular interest in studying polymers because of their structural diversity.

The ability to form network structures constitutes one of the most interesting and useful options in polymer design. Network formation can occur by crosslinking chains of a linear polymer or by curing of multifunctional thermosetting resins. The macroscopic physical properties of the system can be improved considerably in some cases; better temperature and size stability, increased mechanical strength, and improved adhesion are all possible results. As a material becomes more highly crosslinked, however, it becomes increasingly intractable and generally loses solubility. The details of network formation then become difficult to characterize. It is this problem which magic-angle C-13 NMR addresses. By analyzing the carbon resonance behavior, one can determine the reactions which occur during crosslinking. In many cases the interpretation of results is straightforward, as will be demonstrated by discussion of the following systems: a) Crosslinking of elastomers; free-radical crosslinking of polybutadiene and polyisoprene has been known as a means of improving material properties for many years (17,18). Solution-state C-13 NMR has been used to study the mechanisms of such reactions (19,20) and can clearly indicate which carbons are preferentially attacked. However, the resonance broadening which accompanies even modest levels of crosslinking limits the information obtainable from solution work. Cross polarization, magic-angle spectra of cis-polybutadiene cured for 2 hrs at 150°C are shown in Figure 1. The free radicals responsible for crosslinking are generated at various concentrations by the dissociation of dicumyl peroxide. At higher degrees of crosslinking, the resonances broaden appreciably. This indicates a wider distribution of isotropic chemical shifts, arising from the differences in bonding and molecular packing associated with network formation. A second interesting observation is that the total integrated area of the spectrum increases dramatically with higher crosslinking. Cross polarization discriminates heavily against highly mobile regions, because of the intensity dependence upon the static dipolar coupling between carbons and protons. The increase in resonance areas then reflects the higher rigidity of the samples following crosslinking. The final, most significant point is the appearance of the broad quaternary carbon resonance at 45 ppm. This peak arises from the carbons directly attacked by the radicals. Hydrogen abstraction occurs, and crosslinks are formed according to the following mechanism:

CROSS-LINKED POLYMERS: THE FORMATION OF NETWORKS

$$ROOR \rightarrow 2RO$$

$$-(CH_2-CH=CH-CH_2-) + RO \rightarrow$$
$$-(CH-CH=CH-CH_2-)+ROH$$
$$-CH-CH=CH-CH_2-$$
$$2-(CH-CH=CH-CH_2-)\rightarrow \quad |$$
$$-CH-CH=CH-CH_2-$$

The formation of the <u>trans</u> configuration is possible during the rearrangement initiated by radical attack. Spectral evidence of this is seen in the low-field shoulder of the methylene peak at approximately 32 ppm of the spectrum corresponding to 5 phr dicumyl peroxide. A wide range of additional information dealing with network formation in elastomers is obtainable with magic-angle NMR (21,22). Areas of particular interest include differentiating mechanisms in the cure of polybutadiene and polyisoprene and reactions involved in the formation of sulfur vulcanizates.

POLYMER CRYSTAL
STRUCTURES

X-ray and electron diffraction are the most common means of analyzing the crystal structures of polymers. For certain systems, however, these techniques encounter difficulties because of lack of perfect ordering of the polymer chains or because of incorrect assumptions of the monomeric structure in the unit cell. The high-resolution C-13 spectra obtained with magic-angle spinning and cross polarization can yield additional information which may help to resolve such difficulties. Although they do not give direct structures, the spectra indicate the number of inequivalent nuclei in the unit cell through the splitting pattern of the resonances. The most detailed NMR work to date in this area has been done on cellulose (23-25). Resonances corresponding to the two carbons (C-1 and C-4) which support the glycosidic linkages between adjacent anhydroglucose units are split. The pattern suggests that four inequivalent anhydroglucose units and two different kinds of glycosidic linkages are present in the unit cell of cellulose I. This is in disagreement with currently accepted X-ray structures (26-28), which call for two magnetically inequivalent anhydroglucose monomers per unit cell. The NMR evidence led Atalla et al. (23) to suggest that the X-ray structure determinations should be performed using anhydrocellobiose as the basic repeat unit, instead of anhydroglucose. Earl and VanderHart (25) indicate that relaxation of the symmetry of the twofold screw axis would explain the resonance splitting and would agree in large part with the X-ray work which has been reported. These studies illustrate the specific information for polymeric crystals available from NMR, without the direct determination of structures in terms of bond lengths and angles.

Harris et al. (29) have reported the magic-angle spinning, cross-polarization spectra of poly (1-butene) and syndiotactic polypropylene. Their results indicate again that NMR can be used to differentiate among crystal structures. Poly(1-butene) has a number of crystal forms, one of which is stable at room temperature and another of which is obtained by heating at 130°C. After removal of heat, the second form reverts to the first as a function of time. This transition has been studied by X-ray diffraction (30) and by vibrational spectroscopy (31). The NMR spectra for this process are shown in Figure 2. In A the spectrum of the high-temperature form displays relatively poor sensitivity and broad resonances. As the structure relaxes to the room-temperature form over a period of several days, the linewidth decreases substantially, and the sensitivity improves. Differences in relaxation times between the two crystal forms are probably responsible for the spectral changes, but this has not yet been investigated fully. In the case of polypropylene, the crystal structures of the isotactic and syndiotactic forms can be differentiated easily through their NMR spectra. The isotactic material crystallizes in a simple helical form (32) and has one resonance each for the methyl, methylene, and methine carbons. The syndiotactic polymer has an involuted helical structure in the crystal (33), which generates an inequivalence between successive methylene carbons in the chain. The magic-angle spectrum clearly resolves the two methylene carbons, which are separated by 8 ppm. The magnitude of this effect agrees well with a γ-effect of -4 ppm. For the low-field methylene carbons, both γ-carbons are in the trans conformation, while for the high-field ones, both γ-carbons are in the gauche conformation. The appearance of this additional resonance clearly points out a difference in the crystal structures of the isotactic and syndiotactic materials. While some clues as to the nature of the difference may be obtained from the NMR spectra, diffraction results are necessary for more specific information.

Magic-angle C-13 NMR also has been applied to several small molecules in order to obtain information relevant to crystalline structures (16,34,35). The techniques and interpretations used in this work should be of value in understanding some aspects of polymer crystals.

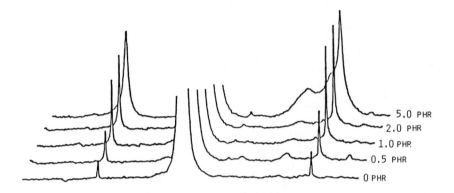

Figure 1. Spectra of cis-polybutadiene cured for 2 h at 149 °C with various levels of dicumyl peroxide (given in parts per hundred at the right). The C-13 spectra were recorded at 38 MHz with cross polarization and magic-angle spinning. The truncated resonance in the center of the spectra arises from the polyoxymethylene rotor.

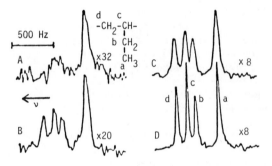

Figure 2. CP/MAS spectra of isotactic poly(1-butene) taken at 22.6 MHz. Samples were heated to form the high-temperature crystal structure and then cooled to room temperature. Spectra were taken at various times after recooling: (A) immediately, (B) after 20 h, (C) after 2 d, and (D) after 13 d. Vertical scaling factors are indicated. (Reproduced with permission from Ref. 29. Copyright 1981, Hüthig and Wepf Verlag, Basel.)

IR spectroscopy is an old and familiar technique for polymer characterization. It is base on the absorption of radiation in the IR frequency range due to the molecular vibrations of the functional groups contained in the polymer chain. Prior to FTIR, IR spectroscopy was carried out using a dispersive instrument utilizing prisms or gratings to geometrically disperse the infrared radiation (Figure 3). Using a scanning mechanism, the dispersed radiation was passed over a slit system which isolated the frequency range falling on the detector. In this manner, the spectrum, that is, the energy transmitted through a sample as a function of frequency, was obtained. This infrared method is highly limited in sensitivity because most of the available energy is being thrown away, i.e. does not fall on the open slits. Hence, to improve the sensitivity of infrared spectroscopy, a technique is sought which allows the examination of all of the transmitted energy all of the time.

The Michelson interferometer is such an optical device (Figure 4). The Michelson interferometer has two mutually perpendicular arms. One arm of the interferometer contains a stationary, plane mirror; the other arm contains a movable mirror. Bisecting the two arms is a beam splitter which splits the source beam into two equal beams. These two light beams travel down their respective arms of the interferometer and are reflected back to the beamsplitter and on to the detector. The two reunited beams will interfere constructively or destructively, depending on the relationship between their path difference (x) and the wavelengths of light. When the movable mirror and the stationary mirror are positioned the same distance from the beamsplitter in their respective arms of the interferometer (x=0), the paths of light beams are identical. Under these conditions all wavelengths of the radiation striking the beam splitter after reflection add coherently to produce a maximum flux at the detector and generate what is known as the "center burst". As the movable mirror is displaced from this point, the path length in that arm of the interferometer is changed. This difference in path length causes each wavelength of source radiation to destructively interfere with itself at the beam splitter. The resulting flux at the detector, which is the sum of the fluxes for each of the individual wavelengths, rapidly decreases with mirror displacement. By sampling the flux at the detector, one obtains an interferogram. For a monochromatic source of frequency v, the interferogram is given by the expression

$$I(x) = 2\ R\ T\ I(v)(1 + \cos 2\pi\ vx)$$

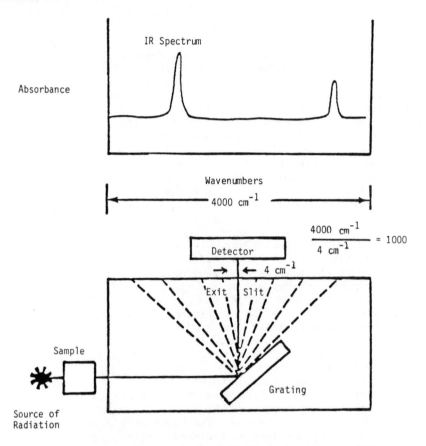

Figure 3. *Optical diagram of the IR dispersive spectrometer.*

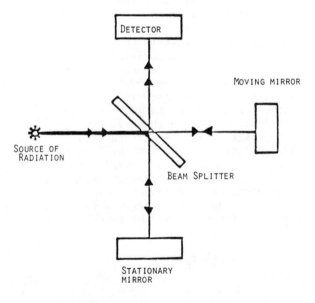

Figure 4. *Optical diagram of the interferometer.*

where R is the reflectance of the beam splitter, T is the transmittance of the beam splitter, I(v) is the input energy at frequency v and x is the path difference. The interoferogram consists of two parts, a constant (DC) component equal to 2 R T I, and a modulated (AC) component. The AC component is called the interferogram and is given by:

$$I(x) = 2\ R\ T\ I(v)\ \cos(2\pi\ vx)$$

An infrared detector and an AC-amplifier convert this flux into an electrical signal:

$$I(x) = re\ I(x)\ volts$$

where re is the response of the detector and amplifier.

The moving mirror is generally driven on an air bearing and for FTIR measurements the mirror must be kept in the same plane to better than 10 micron-radians for mirror drives up to 10 cm in length. It is necessary to have some type of marker to initiate data acquisition at precisely the same mirror displacement everytime. The uncertainty in this position cannot be greater than 0.1 microns from scan to scan. This precise positioning is accomplished by having on the moving mirror a smaller reference interferometer. Through this reference interferometer is passed a visible white light source. This white light source has a very sharp "center burst" or spike and the data acquisition is initiated when this signal reaches a predetermined value.

Digitization of the interferogram at precisely spaced intervals is accomplished with the help of the interference signal of the auxilliary interferometer equipped with a He-Ne laser which yields a monochromatic signal resulting in the output of a cosine function from the interferometer (Figure 5). The interferogram is sampled at each zero-crossing of the laser cosine function. Therefore, the path difference between two successive data points in the digitized interferogram is always a multiple of half wavelengths of the laser, or 0.316 microns. The laser also provides an internal calibration of the wavelength. The interferogram for a polychromatic source A(v) is given by

$$I(x) = \int_0^\infty A(v)\ (1+\cos\ 2\pi vx)dv$$

The methods of evaluating these integrals involve a determination of the values at zero-path length and very long or infinite path length. At zero difference:

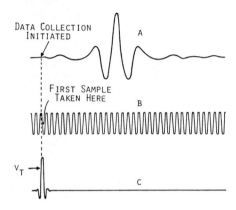

Figure 5. Signals from an interferometer with a separate reference cube for measuring the laser and white light interferograms. Key: A, signal (IR) interferogram; B, laser interferogram; and C, white light interferogram. The zero retardation position of the reference interferometer has been displayed relative to that of the main interferometer so that data collection may be initiated a short distance before the maximum of the IR interferogram. (Reproduced with permission from Ref. 79. Copyright 1978, Plenum Publishing Corp.)

$$I(o) = 2 \int_0^\infty A(v)dv$$

and for large path differences

$$I(\infty) = \int_0^\infty A(v)dv = I(o)/2$$

so the actual interferogram $F(x)$ is:

$$F(x) = I(o)-I(\infty)=\int_0^\infty A(v) \cos (2\pi vx) dv.$$

From Fourier transform theory (38,39,40,41,42).

$$A(v) = 2 \int_{-o}^{+\infty} F(x) \cos (2\pi vx)dx$$

This Fourier transform process was well known to Michelson and his peers but the computational difficulty of making the transformation prevented the application of this powerful interferometric technique to spectroscopy. An important advance was made with the discovery of the fast Fourier transform algorithm by Cooley and Tukey (43) which revived the field of spectroscopy using interferometers by allowing the calculation of the Fourier transform to be carried out rapidly.

The typical FTIR instrument has a number of components including a minicomputer, a moving head disc, ADC capable of operating up to 50 kHz, input—

output interfaces, data terminal for input, a high
speed digital plotter, and an oscilloscope with
graphic and alphanumeric display. For software, all
commercial systems have complete data processing
capabilities including signal averaging, apodization,
phase correction, gain-ranging and a variety of
programs to manipulate and display the spectral data.
With every month, new software programs are being
developed and exchanged between the users of the
equipment. A number of companies offer turn-key
instruments with a wide range of resolution, sensitivity, frequency range and speed. With the large
commercial market for FTIR, continued improvement
in hardware and software can be expected. The
sampling accessories for FTIR are shown in Figure 6
and cover the complete range of techniques.

The advantages of FTIR over dispersive
(i.e. grating) infrared arise from several sources.
Fellgett's advantage (44) or the multiplex advantage is the principle advantage of FTIR. For measurements taken at equal resolution and for equal
measurement time with the same detector and optical
throughput, the signal-to-noise (S/N) of spectra from
an FTIR will be M times greater than on a grating
instrument where M is the number of resolution
elements being examined during the measurement.
Alternately, for a given observation time, it is
possible to repeat the FTIR measurement M times
which increases the signal by a factor of M and the
noise by a factor of $M^{\frac{1}{2}}$, to achieve a S/N enhancement of a factor of $M^{\frac{1}{2}}$. This advantage arises from
the fact that the FTIR spectrometer examines the
entire spectrum in the same period of time required
for a dispersive instrument to examine a single
spectral element. Theoretically, an FTIR instrument
can acquire the spectrum with the same S/N from 0
to 4000 cm^{-1} with 1 cm^{-1} resolution 4000 times faster
than a dispersive instrument. Or from another point
of view, for the same measurement time a factor of
approximately 63 increase in S/N can be achieved on
the FTIR instrument. Therefore, when there is a
limited time for measurement, there is a definite
time advantage for the FTIR instrument. When time
of measurement is not important consideration, the
time can be used to multiscan with the FTIR instrument to signal average and increase the S/N. Of
course, there is also the inherent time advantage
associated with rapid scanning FTIR since it requires
a very short time to scan the mirror and obtain the
complete spectra (ca. 1.5 sec). This time advantage
of the FTIR has been particularly important for the
study of polymerization chemistry (45), degradation

COMPARISON OF FTIR
WITH DISPERSIVE IR
SPECTROSCOPY

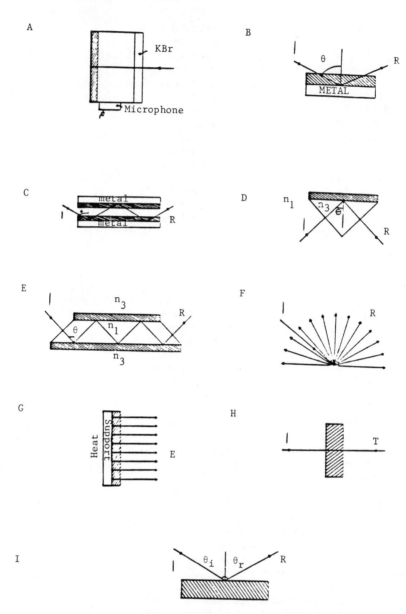

Figure 6. (A) PAS cell. The incident light produces pressure fluctuations that are detected by a sensitive microphone. (B) Single reflection RA set-up. Light penetrates the sample first and is reflected by the metal mirrors (θ should be 70° to 89.5°). (C) Multiple reflection RA set-up. Light penetrates the sample first and is reflected by the metal mirrors (θ should be 70° to 89.5°). (D) Single reflection IRS set-up. Light passes through the IRE first and is totally reflected at $\theta > \theta_c$: $n_1 \sin \theta_c = n \sin 90°$; $\sin \theta_c = n_2/n_1$. (E) Multiple reflection IRS set-up. (F) Diffuse reflectance. The scattered light is collected by mirrors and directed to the detector. (G) Emission technique. The sample is heated and the emitted radiation is analyzed. (H) Transmission spectroscopy. (I) Spectral reflection (mirror-like). The angle of incidence equals the angle of reflection.

processes (46) and other time-dependant processes of polymers (47,48) to be discussed later.

The Jacquinot or throughput advantage (49) arises from the loss of energy in the dispersive system due to the gratings and slits. These losses do not occur in an FTIR instrument which does not contain these elements. Basically, the Jacquinot advantage means that the radiant power of the source is more effectively utilized in interferometers. The throughput for an FTIR instrument is limited by the size of the mirrors. The Jacquinot's advantage compared for an interferometer and a commercial dispersive spectrometer has been given (50). This higher throughput is particularly important in the infrared region where the signals are weak since the infrared sources are weak. The throughput advantage has been used for studying strongly absorbing systems such as carbon-black filled rubbers (51) and emission from polymers (52).

The far infrared region is difficult to study due to energy limitations. The commonly used blackbody sources, such as Nernst glowers or globars contribute less than one hundredth of one percent of the total energy in the region below 100 cm^{-1} with peak output near 3000 cm^{-1}. An additional problem in the far infrared is the elimination of the unwanted energy from shorter wavelengths. Finally, detectors for the long wavelength region are poor so the overall sensitivity in the far-infrared region is further reduced. But with FTIR, this region is accessible for study and a number of polymers have been examined in this region as will be discussed later.

The combination of Fellgett's and Jacquinot's advantage coupled with the inherent speed differential should lead to an enormous difference between FTIR and dispersive instruments. However, in practice, part of this advantage is offset by the difference in the performance of the triglycine sulfate (TGS) and thermocouple detectors. At low modulation frequencies, the thermocouple detector is about an order of magnitude more sensitive than TGS.

The Conne (53) or frequency advantage comes from the fact that the frequencies of an FTIR instrument are internally calibrated by a laser whereas conventional IR instruments exhibit drifts when changes in alignment occur. This latter advantage is particularly useful for coaddition of spectra to signal average since the frequency accuracy is an absolute requirement in this case. For the absorbance subtraction technique to be useful for samples examined over a period of time such as months or years, frequency accuracy must be maintained. Thus with FTIR, polymer samples can be scanned, the spectra stored, and years later compared with spectra of the samples run currently. Applications such as quality control and long term aging and weathering immediately come to mind based on the reproducibility of the frequency of an FTIR instrument over the long term.

Another area of superiority of interferometers is the ease with which stray light is reduced. For a grating monochromator, the stray light arises in two different ways. The largest source occurs when radiation diffracted from one order of a grating reaches the exit slits, radiation reflected from other orders is also present. The other source of stray light arises from reflection in the monochromator of nondispersed light which can reach the detector. There is no way of distinguishing this flux of scattered light from the desired flux so the dynamic range of the detector is decreased. This problem can be severe particularly in energy limited situations. With an interferometer, stray light of the second type which has not been modulated by the interferometer appears at the detector as a constant DC offset, and, as such, has no effect on the computed spectrum. Efficient electronic filtering of the signal which avoids folding effects eliminates this unwanted flux contribution. While there is not direct correspondence to overlapping orders in an interferometer, stray light which has been modulated is also easily handled. If the interferogram is sampled at sufficiently high frequency, energy below this Nyquist frequency will be properly accounted for in the computed spectrum. Since the Fourier components due to high frequency radiation vary most rapidly, it is possible to eliminate them from the interferogram by the inclusion of a simple audio band-pass filter. This type of filtering has the advantage that it occurs after the detector and does not effect the light throughput.

The overall simplicity of an FTIR compared to a dispersive instrument is also an advantage. For example, a single instrument can be easily converted to study the near, mid or far-infrared frequency region whereas with the dispersive method, three totally different instruments are required. To improve resolution with an FTIR instrument, the basic design is only slightly modified while for a dispersive instrument different optical components are required.

DATA PROCESSING
TECHNIQUES USING
DIGITIZED IR SPECTRA

Absorbance Subtraction. One of the spectral processing operations most widely used in polymer analysis is the digital subtraction of absorbance spectra in order to reveal or emphasize subtle differences between two samples or a sample and a reference material (36). The number of polymer applications of absorbance subtraction is rapidly increasing and it is this digital subtraction capability more than any other single factor that has inspired subsequent investigation of polymeric materials using FTIR. Spectral subtraction is a powerful method of extracting structural information about components of composite spectra (37). When a polymer is examined before and after a chemical or physical treatment, and subtraction of the original spectrum from the final spectrum is done, positive absorbances reflect the structures that are formed during the treatment and negative absorbances reflect the loss of structure (54). The advantage of FTIR difference spectra lies in the ability to compensate for differences in thicknesses of the two samples. This balance of thicknesses allows small spectral differences to be associated with structural changes and not to be outweighed by the differences in the amount of sample in the beams. Additionally, with properly compensated thicknesses, the differences in absorbances can be magnified through computer scale expansion to reveal small details of the spectral differences. The scaling parameter, k, is chosen such that

$$(A_1 - k A_2) = 0$$

where A_1 and A_2 correspond to the absorbances of the internal thickness bands of samples one and two. Multiplication of the spectrum of sample 2 by k will yield a new spectrum having the same optical thickness as sample 1. One may use the peak absorbances, integrated peak areas, or a least-squares curve fitting method to calculate the scaling factor k. The method of choice will depend on the system being examined. It should be remembered that the calculation of the scaling factor cannot be done entirely analytically and the only test of the scaling factor is the resultant difference spectrum, which should be examined carefully before any further analysis is carried out.

In Figure 7, the absorbance spectrum is shown for a polybutadiene sample which has undergone only slight oxidation. The difference spectrum expanded 100 times shows the initial sites of oxidation.

Absorbance subtraction can be considered as a spectroscopic separation technique for some problems in polymers. An interesting application in FTIR difference spectroscopy is the spectral separation of a composite spectrum of a heterophase system. One such example is a semicrystalline polymer which may be viewed as a composite system containing an amorphous and crystalline phase (54). In general, the infrared spectrum of each of these phases will be different because in the crystalline phase one

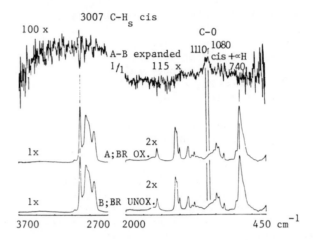

Figure 7. cis-1,4-Polybutadiene oxidation, 1 h at 25 °C. Bottom, unoxidized; center, oxidized; top, difference spectrum. (Reproduced from Ref. 80. Copyright 1976, American Chemical Society.)

particular rotational conformation will predominate whereas in the disordered amorphous regions a different rotamer will dominate. Since the infrared spectrum is sensitive to conformations of the backbone, the spectral contributions will be different if they can be isolated. The total absorbance A_t at a frequency v of a semi-crystalline polymer may be decomposed into the following contributions:

$$A_t (v) = A_c(v) + A_a(v) + A_i(v)$$

where $A_c(v)$ and $A_a(v)$ are the contributions to the total absorbance at frequency v due to the crystalline and amorphous components, respectively, and $A_i(v)$ is the contribution to the total absorbance at frequency v due to phase independent absorptions. For the sample of higher crystallinity

$$A_{1t}(v) = A_{1c}(v) + A_{1a}(v) + A_{1i}(v)$$

while for the lower crystallinity sample one can write a similar equation. Once assignments have been made as to the amorphous and crystalline nature of a band in the spectrum by systematic annealing studies, the two spectra can, with a suitable choice of scaling parameters, be subtracted from each other until bands assigned to one of the phases have been reduced to the background. To generate the pure

spectra of the crystalline component one subtracts
the spectra of the two samples thusly,

$$(A_t - k A_{2t}) = (A_{1c} - k A_{2c}) + (A_{1a} - k A_{2a}) +$$

$$(A_{1i} - k A_{2i})$$

The resultant difference spectrum is the "purified"
spectrum of the crystalline phase. The method is
illustrated for polystyrene in Figure 8. A similar
set of equations holds for generating the amorphous
phase spectrum. This technique can be applied to
other separations where a variation in the relative
amount of the structural components can be done.

This technique was first applied to determine
the crystalline vibrational bands of trans-1,4-
polychloroprene (55). The spectrum of a cast film
of predominantly (90%) trans-1,4-polychloroprene
polymerized at -20 °C was compared with the same
sample heated to 80 °C (above the melting point) for
15 min (Figure 9). Elimination of the amorphous
contribution of the composite semicrystalline spectrum was accomplished by subtracting spectrum b
from spectrum a until the bands at 602 and 1227 cm^{-1}
were reduced to the baseline. The "purified" crystalline spectrum is given by spectrum c at the top
of Figure 9, and exhibits the sharp band structure
expected for a regular crystalline array. The interesting aspect of the crystalline spectrum was the

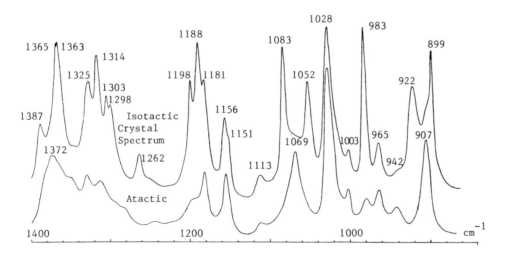

Figure 8. Comparison of the crystal difference spectrum of
IPS with the IR spectrum of an atactic sample. (Reproduced
with permission from Ref. 58. Copyright 1977, John Wiley &
Sons, Inc.)

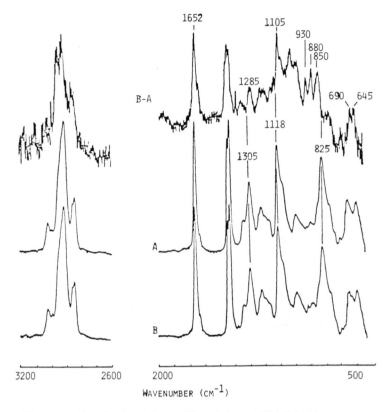

Figure 9. Fourier transform IR spectra at 70 °C in the range 500-3200 cm⁻¹. Key: A, polychloroprene polymerized at -20 °C; B, polychloroprene polymerized at -40 °C; B-A, difference spectrum. (Reproduced with permission from Ref. 76. Copyright 1978, Butterworth & Co., Ltd.)

observation that when the crystalline component spectra were obtained for samples polymerized at different temperatures, through the same procedure, the crystalline vibrational frequencies were different (56). This should not be the case if the crystalline phase had the same structure. However, the spectra indicated that structural defects were being imbibed into the crystalline domains and there were more defects as the polymerization temperature increased. As a result of the defects occuring in the crystalline phase, the structure of the crystalline phase was different and consequently the spectra reflected this difference. However, to date, this is the only crystalline spectrum showing this effect, yet it was the first observed!!

The separation of the crystalline and amorphous

phases into their respective spectra has been carried
out for a number of polymers including polyethylene
terephthalate (57), polystyrene (58), poly(vinyl
chloride) (59), polyethylne (60,61), nylon (62),
polypropylene (63), and poly(vinylidene fluoride) (64).

Ratio Method. One of the major problems in studying polymers quantitatively is the absence of
model compounds for the purpose of calibration. A
method of obtaining spectra of the components of a
mixture spectra is based on obtaining the ratio of
absorbances. This method was first used by
Hirschfeld (65) for mixtures of components differing
in relative concentration. This approach was later
generalized but is limited to a rather small number
of components since otherwise it is difficult to
sort out the various ratios associated with each
component (66). Even more recently a careful look
has been taken at the limitations of the absorbance
ratio method (67).
    The infrared spectrum of a two-component mixture
can be represented by

$$M(v) = f_1(v) + f_2(v)$$

where $M(v)$ = spectrum of a mixture of components and
$f_1(v)$ = spectrum of a pure component i. The spectrum
of a mixture of the same two components in different
proportions is represented as

$$M_1(v) = a_1 f_1(v) + a_2 f_2(v)$$

Solving the equations for $f_1(v)$ and $f_2(v)$ one
obtains

$$f_1(v) = \frac{1}{a_1 - a_2} M(v) - \frac{a_2}{a_1 - a_2} M(v)$$

$$f_2(v) = \frac{1}{a_2 - a_1} M_1(v) - \frac{a_1}{a_2 - a_1} M(v).$$

The ratio spectrum

$$R(v) = \frac{M_1(v)}{M(v)} = \frac{a_1 f_1(v) + a_2 f_2(v)}{f_1(v) + f_2(v)}$$

defines the coefficients a_1 and a_2 by means of "flat
areas". In a spectral region where $f_1(v) \gg f_2(v)$,
$R(v) \approx a_1$. Conversely, if $f_2(v) \gg f_1(v)$ in any
region, $r(v) \approx a_2$. The accuracy to which $f_1(v)$ and

$f_2(v)$, the pure component spectra, are calculated depends solely on the accuracy to which a_1 and a_2 are determined. Computer simulation studies indicate that the determination of the coefficients is inaccurate if band overlap occurs or if small frequency shifts occur with the differing concentrations in the mixtures. The advisable method is to seek the maxima and minima in the spectra and calculate the spectra. Mixing of the spectra will be revealed by apparently negative absorbances in each of the component spectra. It is then necessary to iterate the values of the coefficients until the calculated spectra are "unmixed". It is to be observed that the calculated spectra are not the spectra normalized to unit concentrations or molarity, but contain a relative concentration term which depends on the concentrations of the initial mixtures used to calculate the spectra. It is necessary to carry out an "internal calibration" to obtain the specific or molar absorptivities (68). This method has proved feasible in deriving spectra of the pure components from the spectra of two component mixtures. Care must be taken to properly account for band overlap and band shift problems. The method has the power to calculate spectra of "pure" component spectra for systems where the "pure" components cannot be modeled by suitable standard compounds. In particular, for polymer applications the spectra of a "pure" crystalline and a "pure" amorphous polymer can be obtained whereas in reality such pure model systems cannot be prepared. It is also possible to calculate the spectra of pure rotational isomers of polymer chains whereas in reality such pure systems have not been prepared (69). This method has been successfully used to obtain the spectra of the pure gauche and trans isomers of poly(ethylene terephthalate) (70) and the spectra are shown in Figures 10 and 11.

LEAST SQUARES CURVE FITTING FOR QUANTITATIVE ANALYSIS

Classically, quantitative infrared analysis was carried out using a single analytical frequency characteristic of the structure being determined in polymers. Least-squares regression analyses have been developed for utilization in the determination of multicomponent mixtures with overlapping spectral features (71). A least-squares analyses uses all of the spectral data in the region of interest. The inclusion of all of the data in the spectral region substantially improves the precision and accuracy of the results (72). The newer programs allow the simultaneous determination of the baselines as well as the spectra. It is also possible to set a

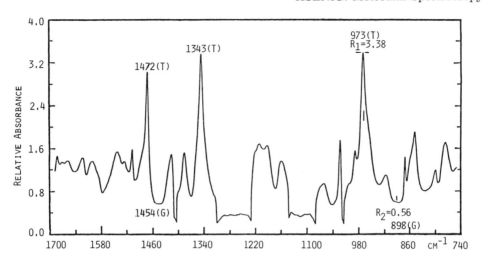

Figure 10. Absorbance ratio spectra of highly annealed quenched PET films.

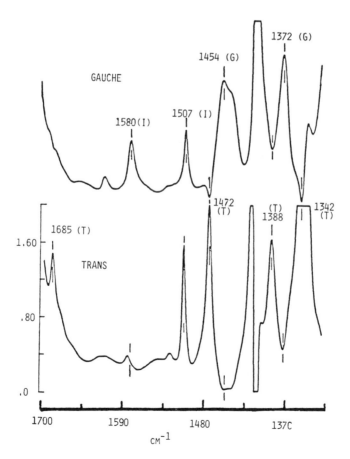

Figure 11. Pure spectra of trans- *and* gauche-*isomers of PET.*

threshold value below which the data will not be used, thus using only those regions of the spectra with spectral information (73). Weighting factors can be used to make maximum use of the spectral data available and minimize those regions with high noise response (73). One of the advantages of the least-squares techniques is that no assumption is made about the spectral line shapes and this aspect of the technique makes it particularly useful for polymer analysis where the band shapes are asymmetric. It is possible to use the least-squares technique to do band shape analysis (74) and to establish non-subjective criteria for absorbance subtraction (75).

SPECTROSCOPIC
TECHNIQUES
USING FTIR

The data manipulating capability of a computerized IR spectrometer allows the spectroscopist to delve more deeply into the structural origin of the infrared absorptions by using data processing techniques to purify, manipulate, and correlate the spectra. If one can systematically vary the relative amounts of various structural contributions, absorbance subtraction can be used to isolate the spectral contributions of the structural components.

The problem encountered in attempting to detect the presence of structural irregularities is the considerable spectral band interference by the strongly absorbing predominant structures. Spectral subtractions allow the removal of these interferences. For example in polychloroprene, the predominant structure is the trans-1,4-polychloroprenes but there exist contributions due to cis-1,4,1,4 and 3,4-structural irregularities depending on the polymerization temperature. Coleman, et al. (76) isolated the spectral contributions of these minor structures by spectrally subtracting out (above the melting point of the polymers) the bands attributable to trans-1,4-units at 1660, 1305, and 825 cm^{-1}. (Figure 9) shows the spectra at 70°C for a polychloroprene polymerized at -20°C (a) and at -40°C (b) and the difference spectrum (A-B). The major bands in the difference spectrum are due to the cis-1,4 unit, (i.e., 1652, 1285, 1105, 850, 690 and 654 cm^{-1}). The resulting difference spectra allowed the detection of cis-1,4-polychloroprene units at the 4% level and at the 1% level for the 1,2- and 3,4-structural units. From the observation of these spectral bands arising from the structural impurities, band assignments could be made and subsequently used for the quantitative structural analysis of polychloroprene.

For many polymers with relatively simple

chemical chain structures, the properties are dependent to a large extent on the conformational or rotational isomeric distribution in the polymer arising
from its process or thermal history. Poly(vinyl
chloride) PVC is such a polymer and it is possible
to use absorbance subtraction to isolate the conformational sequences of PVC by varying the annealing
process (77). A rapidly quenched film was prepared
by quenching into ice water from the melt at 200°C.
Subsequently, the sample was annealed at 80°C cooled
to room temperature and scanned. A similar process
was carried out for annealing temperatures of 100,
120, 140°C on the same film. In the difference
spectra, positive absorbances reflected increases
of the particular conformational sequence while
negative absorbances reflected decreases. The
observed changes were correlated with the theoretical
predictions (78) and assignments made to the various
conformational sequences such as TTTT, TTTG, and
TTGG.

Studies of annealing of polyethylene have revealed new insights into the structural assignments of
some of the amorphous absorptions (60). For example,
a band at 1346 cm^{-1} appears when single crystal mats
are compared with mats which have been quenched and
subsequently annealed as shown in (Figure 12). The
subtraction criterion is the reduction of the amorphous 1368 cm^{-1} band to the baseline. The difference
spectra obtained by subtracting the spectra of
quenched, quenched then annealed for 2 hr, and
quenched then annealed for 74 hrs mats from the spectrum of the original mat are shown in Figure 12. In
all three difference spectra the 1346 cm^{-1} band
remains strongly positive. This observation indicates an absolute decrease in the 1346 cm^{-1} band upon
quenching. Additionally the 1346 cm^{-1} band is not
found in the spectra of extended-chain crystals

with conformation unique to solution-grown single
crystals which is in addition attributed with the
fold surface (60).

For polypropylene, by using the spectrum of an
annealed sample and subtracting it from a quenched
sample it is possible to obtain a difference spectrum characteristic of the amorphous regions of
polypropylene (63). In Figure 13, the difference
spectrum characteristics of the amorphous phase of
the quenched sample (A) is compared with the difference spectrum characteristic of the ordered phase
of an annealed polypropylene sample. The band at
1376 cm^{-1} was used as the criterion for absorbance
subtraction. The difference spectrum of the amorphous phase is very similar to that of the melt.

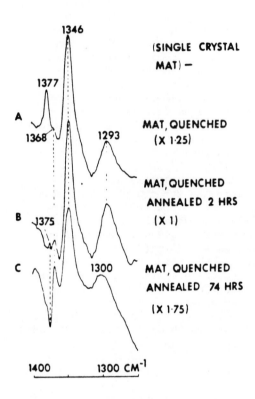

Figure 12. Single crystals: A, quenched, 1:0.65; B, quenched, annealed 2 h, 1:0.95; C, quenched, annealed 74 h, 1:0.98. (Reproduced with permission from Ref. 60. Copyright 1977, John Wiley & Sons, Inc.)

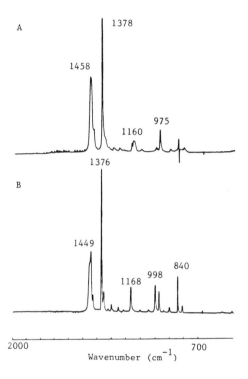

Figure 13. (A) Difference spectrum characteristic of the amorphous phase of the quenched sample. (B) Difference spectrum characteristic of the ordered phase of the annealed sample. (Reproduced with permission from Ref. 63. Copyright 1977, Butterworth & Co., Ltd.)

The features of the spectrum indicate that there are ordered helical chain segments in the amorphous phase of the quenched polypropylene samples.

The spectrum of crystalline isotactic polystyrene has been isolated by subtracting the spectrum of a quenched (amorphous) and annealed semicrystalline film from each other. The criterion for subtraction was the reduction of the amorphous band near 550 cm^{-1} to the baseline (58). It is not precisely correct to term the resultant spectrum a crystalline spectrum since no effects due to interchain interactions have been isolated. It is more properly described as typical of long segments of helical conformations. The vast majority of such chains are found in the crystalline phase (Figure 8). The scale-expanded difference spectrum reveals new features. For the first time the doublets at 1365-1363 and 1303-1298 cm^{-1} are observed. The controversy over the origin of the doublet at 1083-1052 cm^{-1} is cleared up by an examination of a series of difference spectra resulting from annealing. The results suggest that the splitting is associated with the sequence length of the preferred conformation in the amorphous region rather than a rotational disorder of the benzene ring in the crystalline regions.

LITERATURE CITED

1. Stothers, J.B., "Carbon-13 NMR Spectroscopy"; Academic Press: New York, 1972.
2. Wehrli, F.W., and Wirthlin, T., "Interpretation of Carbon-13 NMR"; Heyden & Sons: Philadelphia, 1978.
3. Lyerla, J.R., "High-Resolution Nuclear Magnetic Resonance Spectroscopy", in "Methods of Experimental Physical"; Academic Press, Vol. 16A: New York, 1980, pp. 241-369.
4. McBrierty, V.J., and Douglass, D.C., Physics Reports 1980, 63, 61.
5. McBrierty, V.J., and Douglass, D.C., J. Polym. Sci. Macromolecular Reviews 1981, 16, 295.
6. Bloch, F., Phys. Rev. 1958, 111, 841.
7. Mehring, M., "High Resolution NMR Spectroscopy in Solids"; Springer-Verlag: New York, 1976.
8. Andrew, E.R., Prog. Nucl. Magn. Reson. Spectrosc. 1974, 8, 1.
9. Schaefer, J., and Stejskal, E.O., "High-Resolution C-13 NMR of Solid Polymers", in "Topics in Carbon-13 NMR Spectroscopy"; John Wiley & Sons: New York, 1979, pp. 283-324.
10. Lyerla, J.R., "High Resolution Carbon-13 NMR Studies of Bulk Polymers", in "Contemporary Topics in Polymer Science"; Plenum Publ., Vol.

3; Plenum Publ. Corp.: New York, 1979, pp. 143-213.

11. Earl, W.L., and VanderHart, D.L., Macromolecules 1979, 12 762.

12. Hartmann, S.R., and Hahn, E.L., Phys. Rev. 1962, 128, 2042.

13. Pines, A., Gibby, M.G., and Waugh, J.S., J. Chem. Phys. 1973, 59, 569.

14. Schaefer, J., and Stejskal, E.O., J. Amer. Chem. Soc. 1976, 98, 1031.

15. Stejskal, E.O., Schaefer, J., and Waugh, J.S., J. Magn. Reson. 1977, 28, 105.

16. Balimann, G.E., Groombridge, C.J., Harris, R.K., Packer, K.J., Say, B.J., and Tanner, S.F., Phil. Trans., R. Soc. London, 1981, A 299, 643.

17. Bateman, L., "The Chemistry and Physics of Rub-ber-Like Substances", Chap. 15; Maclaren: London, 1963.

18. Van der Hoff, B.M.E., Ind. Eng. Chem., Prod. Res. Develop. 1963, 2 273.

19. Duch, M.W., and Grant, D.M., Macromolecules 1970, 3, 165.

20. Mochel, V.D., "Carbon-13 NMR of Polymers", in J. Macromol. Sci., Rev. Macromol. Chem., 1972, 8, 289.

21. Patterson, D.J., and Koenig, J.L., unpublished data.

22. Okocha, N.V., and Koenig, J.L., unpublished data.

23. Atalla, R.H., Gast, J.C., Sindorf, D.W., Bartuska, V.J., and Maciel, G.E., J. Amer. Chem. Soc. 1980, 102, 3249.

24. Earl, W.L., and VanderHart, J., Amer. Chem. Soc. 1980, 102, 3251.

25. Earl, W.L., and VanderHart, D.L., Macromolecules 1981, 14, 570.

26. Gardner, K.H., and Blackwell, J., Biopolymers 1974, 13, 1975.

27. Sarko, A., and Muggli, R., Macromolecules 1974, 7, 486.

28. French, A.D., Carbohydr. Res. 1978, 61, 67.

29. Harris, R.K., Packer, K.J., and Say, B.J., Makromol. Chem., Supl., 1981, 4, 117.

30. Turner-Jones, A., Polymer 1966, 1, 23.

31. Cornell, S.W., and Koenig, J.L., J. Polym. Sci. A-2, 1969, 7, 1965.

32. Turner-Jones, A, Aizlewood, J.M., and Beckett, D.R., Makromol. Chem., 1974, 75, 134.

33. Corradini, P, Natta, G., Ganis, P., and Temussi, P.A., J. Polym. Sci., 1967, C 2477.

34. Ripmeester, J.A., Chem. Phys. Lett., 1980, 74, 536.

35. Shiau, W.-I., Duesler, E.N., Paul, I.C., Curtin, D.Y., Blann, W.G., and Fyfe, C.A., J. Amer. Chem. Soc., 1980, 102, 4546.
36. Koenig, J.L., Appl. Spect., 1975, 29, 293.
37. D'Esposito, L, and Koenig, J.L., "Applications of Fourier Transform Infrared to Synthetic Polymers and Biological Macromolecules", in "Fourier Transform Infrared Spectroscopy"; Academic Press: J.R. Ferraro, and L.J. Basilie, Eds., vol. 1, chapter 2, 1978.
38. Champeney, D.C., "Fourier Transforms and Their Physical Applications"; Academic Press: New York, 1973.
39. Bracewell, R., "The Fourier Transform and its Applications"; McGraw-Hill: New York, 1965.
40. Foskett, C., in "Transform Techniques in Chemistry"; Plenum Press: New York, P.R. Griffiths, Ed., chapter 2, 1978.
41. Papoulis, A., in "The Fourier Integral and its Applications" McGraw-Hill: New York, 1962.
42. Mertz, L., "Transformation in Optics" John Wiley & Sons: New York, 1965.
43. Cooley, J.W., and Tukey, J.W., Math. Comp. 1965, 19, 297.
44. Fellget, P.B., J. Phys. Radium 1958, 19, 187-237.
45. Antoon, M.K., and Koenig, J.L., J. Polym. Sci. Polym. Chem. Ed., 1981, 19, 549.
46. Liebman, S.A., Ahlstrom, D.H., and Griffiths, P.R., Appl. Spect., 1976, 30, 355.
47. Holland-Moritz, K., Stach, W., and Hollind-Moritz, I., J. Mol. St. 1980, 60, 1.
48. Holland-Moritz, K., Stach, W., and Holland-Moritz, I., Prog. Coll. and Polym. Sci., 1980, 67, 161.
49. Jacquinot, Rep. Prog. Phys., 1960, 13, 267.
50. Griffiths, P.R., Sloane, H.J., and Hannah, R.W., Appl. Spect., 1977, 31, 485.
51. Hart, W.W., Painter, P.C., Koenig, J.L., and Coleman, M.M., Appl. Spect., 1977, 31, 220.
52. Chase, D.B.., Appl. Spect., 1981, 35, 177.
53. Connes, J., and Connes, P., J. Opt. Soc. Am., 1966, 56, 896.
54. Tabb, D.L., Sevcik, J.J., and Koenig, J.L., J. Polym. Sci., Polym. Phys. Ed., 1975, 13, 815.
55. Coleman, M.M., Painter, P.C., Tabb, D.L., and Koenig, J.L., J. Polym. Sci. Polym. Lett. Ed., 1974, 12, 577.
56. Koenig, J.L., Tabb, D.L., and Coleman, M.M., J. Polym. Sci., Polym. Phys. Ed., 1975, 13, 1145.

57. D'Esposito, L, and Koenig, J.L., _J. Polym. Sci. Polym. Phys. Ed._, 1976, 14, 1731.
58. Painter, P.C., and, Koenig, J.L., _J. Polym. Sci. Polym. Phys. Ed._, 1977, 15, 1885.
59. Tabb, D.L., and Koenig, J.L., _Macromolecules_ 1975, 8, 929.
60. Painter, P.C., Havens, J.R., Hart, W.W., and Koenig, J.L., _J. Polym. Sci., Polym. Phys. Ed._, 1977, 15, 1223.
61. Painter, P.C., Havens, J.R., Hart, W.W., and Koenig, J.L., _J. Polym. Sci., Polym. Phys. Ed._, 1977, 15, 1235.
62. Vasko, P.V., and Koenig, J.L., unpublished data.
63. Painter, P.C., Watzek, M., and Koenig, J.L., _Polymer_, 1977, 18, 1169.
64. Bachman, M.A., Gordon, W.L., Koenig, J.L., and Lando, J.B., _J. Appl. Phys._, 1979, 50, 6106.
65. Hirschfield, T.B., _Anal. Chem._, 1976, 4, 721.
66. Koenig, J.L., D'Esposito, L., and Antoon, M.K., _Appl. Spect._, 1977, 31, 292.
67. Diem, H., and Krimm, S., _Proc. from FACS meeting_ Philadelphia, PA., 1980.
68. Koenig, J.L., and Kormos, D., _Appl. Spect._, 1979, 33, 351.
69. Koenig, J.L., and Kormos, D., _Contem. Topics in Polym. Sci._, 1979, 3, 1278.
70. Lin, S.B., and Koenig, J.L., _J. Polym. Sci., Polym. Phys. Ed._, 1982, 20, 2277.
71. Antoon, M.K., Koenig, J.H., and Koenig, J.L., _Appl. Spect._, 1977, 31, 518.
72. Haaland, D.M., and Easterling, R.G., _Appl. Spect._, 1980, 34, 539.
73. Haaland, D.M., and Easterling, R.G., _Appl. Spect._, 1982, 36, 665.
74. Gillette, P.C., Lando, J.B., and Koenig, J.L., _Appl. Spect._, 1982, 36, 401.
75. Gillette, P.C., and Koenig, J.L., _Appl. Spect._ 1982, 36, 661.
76. Coleman, M.M., Petcavich, R.J., and Painter, P.C., _Polymer_ 1978, 19, 1243.
77. Koenig, J.L., and Antoon, M.K., _J. Polym. Sci. Polym. Phys. Ed._, 1977, 15, 1379.
78. Rubcic, A., and Zerbi, G., _Macromolecules_ 1974, 7, 754.
79. Griffiths, P.R. In "Transform Techniques in Chemistry"; Griffiths, P.R., Ed.; Plenum: New York, 1978; 125.
80. Pecsok, R.L.; Painter, P.C.; Shelton, J.R.; Koenig, J.L. _Rubber Chem. Technol._ 1976, 49(4).

Index

UNIVERSITY of STRATHCLYDE
FLECK
18 OCT 1985
LIBRARY
LIBRARIES

Production by Paula Bérard
Indexing by Deborah Corson
Book and cover design by Anne G. Bigler

Elements typeset by Hot Type Ltd., Washington, D.C.
Printed and bound by Maple Press Co., York, Pa.

UNIVERSITY of STRATHCLYDE
FLECK
18 OCT 1985
LIBRARY
LIBRARIES